Food Science and Technology

Food Science and Technology

Ralph Hawkins

STATES
ACADEMIC PRESS
www.statesacademicpress.com

Food Science and Technology
Ralph Hawkins
ISBN: 978-1-63989-209-9 (Hardback)

STATES ACADEMIC PRESS

Published by States Academic Press,
109 South 5th Street,
Brooklyn, NY 11249, USA

Cataloging-in-Publication Data

Food science and technology / Ralph Hawkins.
p. cm.
Includes bibliographical references and index.
ISBN 978-1-63989-209-9
1. Food. 2. Food industry and trade. 3. Food industry and trade--Appropriate technology.
I. Hawkins, Ralph.
TX353 .F66 2022
641.3--dc23

Table of Contents

Preface

The branch of basic and applied sciences which deals with the physical, chemical and biological makeup of food is termed as food science. Its key focus is on studying the mechanisms underlying food deterioration and the processes related to food processing. Food science is a multidisciplinary field which makes use of concepts from the fields of chemical engineering, chemistry, biochemistry and microbiology. It aims to improve the safety, nutrition and availability of food. Food technology refers to the application of principles of food science for preservation, selection, packaging and processing of food. The discipline of food science and technology can be further classified into food chemistry, food engineering, food microbiology, molecular gastronomy and sensory analysis. This book is a compilation of chapters that discuss the most vital concepts in the field of food science and technology. Different approaches, evaluations and methodologies on food science and technology have been included herein. Through this book, we attempt to further enlighten the readers about the new concepts in this field.

To facilitate a deeper understanding of the contents of this book a short introduction of every chapter is written below:

Chapter 1- The multi-disciplinary field involving nutrition, chemistry and biochemistry used in the development of food technology is known as food science. This is an introductory chapter that will explain briefly all the significant aspects of food science as well as food chemistry and food microbiology.

Chapter 2- The branch of science which deals with the production, processing, preservation, packaging, etc. of food is termed as food technology. This chapter has been carefully written to provide an easy understanding of the varied facets of food processing such as stages and methods of food processing, canning and bottling of foods, etc.

Chapter 3- Technology plays a key role in the processing of milk and manufacturing of milk products. Some of the processes which fall under this category are ultra high temperature processing of milk and milk products, and freezing of desserts. The diverse applications of technology in the production of ice cream, dried milk and cheese have been thoroughly discussed in this chapter.

Chapter 4- Technology used in making processed or packaged foods is known as the technology of processed foods. Such foods include cakes and biscuits. Cakes can be processed using the blending method and boiled method. This chapter discusses in detail the technology associated with the production of these types of processed foods.

Chapter 5- The technology that is involved in the processing of meat products is termed as meat processing technology. It includes meat tenderization, cooking and canning of meat. The topics elaborated in this chapter will help in gaining a better perspective about the technology of meat products as well as the effects of processing and preservation technologies on meat.

Chapter 6- Dried seeds, roots or vegetables added to a food item in order to add flavor to it are termed as spices. Condiments constitute sauces, spices or any other preparation to complement a dish or enhance its flavor. This chapter discusses in detail the technologies related to spices and condiments as well as their chemical composition.

I would like to share the credit of this book with my editorial team who worked tirelessly on this book. I owe the completion of this book to the never-ending support of my family, who supported me throughout the project.

Ralph Hawkins

1

Introduction to Food Science

The multi-disciplinary field involving nutrition, chemistry and biochemistry used in the development of food technology is known as food science. This is an introductory chapter that will explain briefly all the significant aspects of food science as well as food chemistry and food microbiology.

Food chemistry is the study of the physical, biological (including microbiological) and chemical makeup of food. Food scientists and technologists apply many scientific disciplines, including chemistry, engineering, microbiology, epidemiology, nutrition and molecular biology to the study of food to improve the safety, nutrition, wholesomeness and availability of food.

Depending on their area of specialization, food scientists may develop ways to process, preserve, package and/or store food, according to industry and government specifications and regulations. Consumers seldom think of the vast array of foods and the research and development that has resulted in the means to deliver tasty, nutritious, safe and convenient foods.

Food Chemistry

Food chemistry deals with the production, processing, distribution, preparation, evaluation, and utilization of food. Food chemists work with plants that have been harvested for food, and animals that have been slaughtered for food. Food chemists are concerned with how these food products are processed, prepared, and distributed. For example, to address consumer demands, some food chemists are involved with finding fat and sugar substitutes that do not alter food taste and texture. Basic food chemistry deals with the three primary components in food: carbohydrates, lipids and proteins.

Carbohydrates make up a group of chemical compounds found in plant and animal cells. They have an empirical formula $C_nH_{2n}O_n$ or $(CH_2O)_n$. Since this formula is essentially a combination of carbon and water these materials are called "hydrates of carbon or carbohydrates". Carbohydrates are the primary product of plant photosynthesis, and are consumed as fuel by plants and animals. Food carbohydrates include the simple carbohydrates (sugars) and complex carbohydrates (starches and fiber).

Lipids include fats, oils, waxes, and cholesterol. In the body, fat serves as a source of energy, a thermal insulator, and a cushion around organs; and it is an important component of the cell. Since fats have 2.25 times the energy content of carbohydrates and proteins, most people try to limit their intake of dietary fat to avoid becoming overweight. In most instances, fats are from animal products – meats, milk products, eggs, and seafood and oils are from plants – nuts, olives, and seeds. We use lipids for flavor, to cook foods, and to improve the texture of foods.

Proteins are important components of food. Every cell requires protein for structure and function. Proteins are complex polymers composed of amino acids. There are 20 amino acids found in the body. Eight of these are essential for adults and children, and nine are essential for infants. Essential means that we cannot synthesize them in large enough quantities for growth and repair of our bodies, and therefore, they must be included in our diet. Proteins consist of long chains of 100-500 amino acids that form into three-dimensional structures, their native state. When you change the native state of the protein, you change the three-dimensional structure, which is referred to as denaturation. Factors that cause denaturation include heating, acid, beating and freezing.

Related Terms in Food Chemistry

- Amino acids: Contain carbon, hydrogen, nitrogen and sometimes sulfur and serve as the monomers to make peptides and proteins. Amino acids have a basic structure that includes an amino group (NH_2) and a carboxyl group (COOH) attached to a carbon atom. There are 20 amino acids found in the body.
- Amylase: An enzyme (protein) in saliva that breaks down starch.
- Antioxidant: Substance that prevents or slows down oxidation; inhibits reactions promoted by oxygen; often used as a preservative.
- Carbohydrate: A compound of carbon and water. Carbohydrates are the most abundant of all carbon-containing compounds, composing nearly three-fourths of the dry mass of all plant life on earth. It is the product of photosynthesis. Examples: glucose, sucrose (table sugar), starch, and cellulose.
- Casein: A protein from milk.
- Coagulation: Transformation of a liquid into a soft or solid mass.
- Denatured: Changed from its natural state. In a denatured protein, its characteristics or properties have been altered in some way, by heat, chemicals, or enzymatic action, resulting in the loss of its biological activity.
- Digestion: The chemical breakdown of large food compounds into smaller molecules that can be absorbed by the intestines in humans and animals. The smaller food molecules travel in the blood and are used by cells to make other components or produce energy needed by the body. Digestion begins in the

mouth as salivary amylase begins to break down starch into simple sugars. It involves the hydrolysis of proteins to amino acids.

- Emulsion: A property where two liquids are evenly spread out in each other, yet not dissolved in each other. Oil and water form the most common emulsions, and milk is an emulsion of butterfat in water. Emulsions are important in the production of foods that contain water and fat, such as mayonnaise or margarine. These products require the addition of an emulsifier, to stabilize food emulsions.
- Enzymatic browning: A biochemical process in which fruit or vegetable tissues turn brown when exposed to oxygen. This process is catalyzed by polyphenol oxidase.
- Enzymes: Protein catalysts, which control specific chemical reactions in living systems (plants and animals). Enzymes are active at low concentrations.
- Ester bonds: Bonds between fatty acids and glycerol molecules.
- Fatty acid: Building blocks of fats, having hydrogen atoms attached to chains of carbon atoms and an oxygen containing carbonyl group (-C=O). Fatty acids are found in every cell of the human body.
- Glycerol: Backbone for fatty acid molecules.
- Glucose: A simple sugar ($C_6H_{12}O_6$) and the primary source of energy for all mammals and many plants. It is also known as dextrose, grape sugar, and corn sugar. It is about half as sweet as table sugar.
- Hormones: Substances that can influence enzyme action, metabolism, and physiology.
- Insoluble: Not capable of being dissolved. Fats are insoluble in water. On the other hand, sugar is soluble in water unless more is added than what a certain volume of water can dissolve, which means that the solvent has become saturated with sugar.
- Lecithin: Emulsifier found in eggs and soybean oil.
- Lipids: Compounds commonly of fatty acids and glycerol. Lipids are the most efficient source of fuel in living things. Food lipids are divided into; 1) fats, which usually come from animal sources and are solid at room temperature; and 2) oils, which usually come from plant sources and are liquid at room temperature. Another type of lipid is cholesterol. Cholesterol is a sterol compound made by animals and is used to make certain steroid hormones in the body. It is not found in plants.
- Melanin: Any of a group of brown or black pigments occurring in plants and animals.

- Organic: Related to the branch of chemistry dealing with carbon compounds. Though all living things contain carbon and thus are considered to be organic, other carbon-containing compounds have been produced in the laboratory.
- Peptide bonds: Covalent bonds between two amino acid molecules.
- Peptides: Short chains of amino acids.
- Photosynthesis: Process by which a plant uses water and carbon dioxide to produce a simple sugar (glucose) and oxygen. Plants store excess sugar as starch.
- Polymers: Contain ten or more monomers. Starch is a polymer of the monomer glucose. A protein is a polymer of amino acids.
- Polyphenol oxidase: A copper-containing enzyme, also called phenolase that catalyzes the oxidation of phenolic compounds contained in plant tissue. Example: it speeds the process of browning of cut apple slices.
- Polyunsaturated: Fatty acids that have multiple double bonds.
- Proteins: Complex polymers composed of amino acid monomers. Some examples of protein are muscle, hair, skin, hormones, and enzymes.
- Rennin: Enzyme used to make cheese.
- Shortening: Crystalline form of a solid fat.
- Soluble: Capable of being dissolved. Gases or solids that dissolve are called solutes, while the liquid that does the dissolving is called the solvent. Like substances are usually soluble in like solvents.
- Starch: A polymer of glucose. It is a complex carbohydrate found in green plants and an important source of energy for animals and humans. During the day, green plants store energy by converting glucose to starch. At night, plants convert starch back to glucose for growth.
- Triacylglycerol: A lipid compound consisting of three fatty acids linked to one glycerol molecule. This compound is an important source of energy for the human body.
- Vegetable oils: Come from plants like corn or soybeans and are an important source of polyunsaturated fats.

Food Microbiology

Food microbiology comprehends the study of microorganisms that colonise, modify, and process, or contaminate and spoil food. It is one of the most diverse research areas within microbiology. It comprises a wide variety of microorganisms including spoilage, probiotic, fermentative, and pathogenic bacteria, moulds, yeasts, viruses, prions, and parasites. It deals with foods and beverages of diverse composition, combining a broad spectrum of environmental factors, which may influence microbial survival and growth. Food microbiology includes microorganisms that have beneficial or deleterious effects on food quality and safety and may therefore be of concern to public health.

The tiniest life forms are bacteria, yeasts, molds, and viruses, termed "microorganisms" because of their size (micro meaning small and organism meaning living being).

Bacteria

Bacteria are the most important microorganisms to the food processor. Most are harmless, many are highly beneficial, some indicate the probable presence of filth, disease organisms, spoilage and a few cause disease. There are thousands of species of bacteria, but all are single-celled and fall into three basic shapes: spherical, straight rods, and spiral rods. To see them, you need a microscope that magnifies about 1000-fold. All bacteria reproduce by dividing into two cells. The two cells then divide to become 4, 4 become 8, and so forth. Under ideal conditions, this doubling may occur as frequently as every 15 minutes, so that within 5 hours there will be more than a million cells from the original single cell. If there are 1000 original cells instead of a single one, there will be over 1 billion cells in 5 hours.

Some rod-shaped bacteria are capable of existing in two forms, dormant spores and active vegetative cells. Vegetative cells form spores under adverse conditions as a means of survival. Spore forms preserve the bacteria from starvation, drying, freezing, chemicals, and heat. When conditions become favorable, the spores germinate, with each spore again becoming a vegetative cell with the ability to reproduce. Among the bacteria, sporulation is not a means of reproduction since each cell forms a single spore which later germinates into a single cell again. Most sporulating bacteria that grow in the presence of air belong to the Genus Bacillus, and most that grow only in the absence of air belong to the Genus Clostridium.

Yeasts and Molds

Yeasts are oval-shaped and slightly larger than bacteria. They reproduce most often by budding. In budding each cell can produce several buds, or swellings, which break away to form new, fully formed daughter cells.

Molds as found on bread, fruit, damp paper, or other surfaces are actually composed of millions of microscopic cells joined together to form chains. The chains usually have numerous branches, called hyphae. Molds can thrive in conditions too adverse for bacteria or yeasts. They reproduce by spores that are frequently present as green or black masses on the protruding hyphae.

Yeasts and molds grow on most foods, on equipment, and building surfaces where there are small amounts of nutrient and moisture. Since bacteria grow faster, they greatly outnumber yeasts and molds in most foods. However, bacteria find conditions of low pH, moisture, or temperature and high salt or sugar unfavorable. In such environments, yeasts or molds predominate. Thus, they can be a problem in dry foods, salted fish, bread, pickles, fruits, jams, jellies, and similar commodities.

Viruses

Viruses are the smallest and simplest microorganisms. Unlike bacteria, yeasts, and molds, viruses are incapable of reproducing independently. Instead, they must first invade the cells of another living organism called the host, before they can multiply. Hence, they are parasitic. Viruses are normally specific in their selection of host cells, some infecting but one species, while others are capable of infecting closely related species. As a result, viruses which infect bacteria, called bacteriophages, cannot infect human beings or other animals. On the other hand, several animal viruses, known as zoonotis, can infect human beings. The viruses are important to the food process in two respects:

- As a bacteriophage of lactic or other fermentative bacteria. Bacteriophage infections of starter cultures can interfere seriously with the manufacture of cheese, buttermilk, sauerkraut, pickles, wine, beer, and other desirable fermentative products.
- As disease transmitted by food to human beings. Although viruses require a live host cell and cannot multiply in foods, they can remain viable and infectious for long periods of time, even under highly adverse conditions, such as drying, freezing, and pasteurization.

Factors Affecting Growth of Microorganisms

The food processor reduces potential problems from microorganisms in several ways:

- Removing or destroying them by trimming, washing, heating, pickling, by adding chemicals, or by encouraging competition by acid- or alcohol-forming organisms.
- Minimizing contamination from equipment, people, the environment, and from unprocessed food.
- Minimizing microbial growth on equipment, by cleaning and sanitizing, and in the product itself by adjusting storage temperature, pH, and other environmental factors.

Although each factor affecting growth is considered separately in the following discussion, these factors occur simultaneously in nature. When more than one condition is somewhat adverse to microbial growth, their inhibitory effects are cumulative.

Temperature

Temperature is the most efficient means to control microbial growth. Based on their tolerance of broad temperature ranges, microorganisms are roughly classified as follows:

- Psychrophies grow only at refrigeration temperatures.
- Psychrotrophs grow well at refrigeration temperatures, but better at room temperature.
- Mesophiles grow best at or near human body temperature, but grow well at room temperature.
- Thermophiles grow only at temperatures about as hot as the human hand can endure, and usually not at all at or below body temperature.

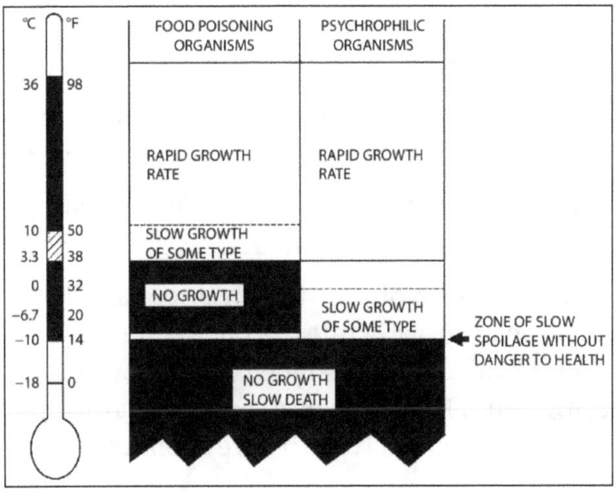

Low temperature limits growth of food poisoning and food spoilage organisms.

To be more specific about these temperature limits of growth is to enter the controversy that has continued since the infancy of microbiology, for there are many species that grow in temperature ranges overlapping these. However, for food microbiology these conclusions are pertinent:

- Some psychrotrophic microorganisms grow very slowly in foods below freezing, but usually not below 19 °F. There are a few reports of growth, usually of molds, at 14 °F, but no reliable reports of growth below that temperature. This means that the standard storage temperature for frozen foods, O °F, does not permit microbial growth. However, many microorganisms survive freezing.
- Most psychrotrophs have difficulty growing above 90 °F.
- Most foodborne disease organisms are mesophiles. The food processor can feel safe in the knowledge that foods held above or below the limits in above figure and rotated properly will remain safe. A good rule of thumb is to store perishable foods below 40 °F or above 140 °F.

- In the temperature range where both mesophilic and psychrotrophic organisms grow (about 41 °F. to about 90 °F), the psychrotrophs grow more rapidly, causing spoilage and at the same time frequently interfering with the growth of foodborne disease organisms.

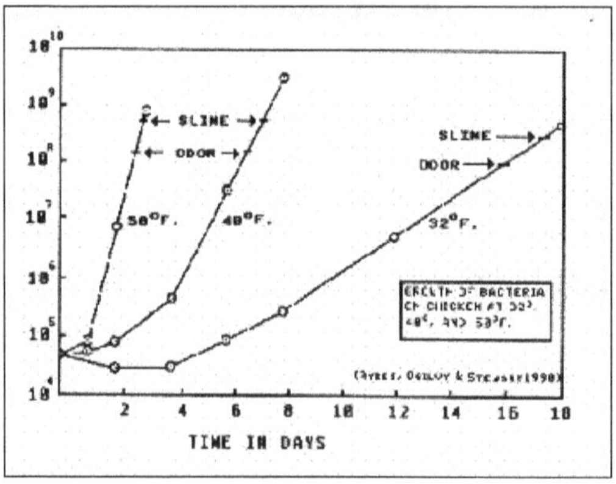

Growth of bacteria on chicken at three temperatures.

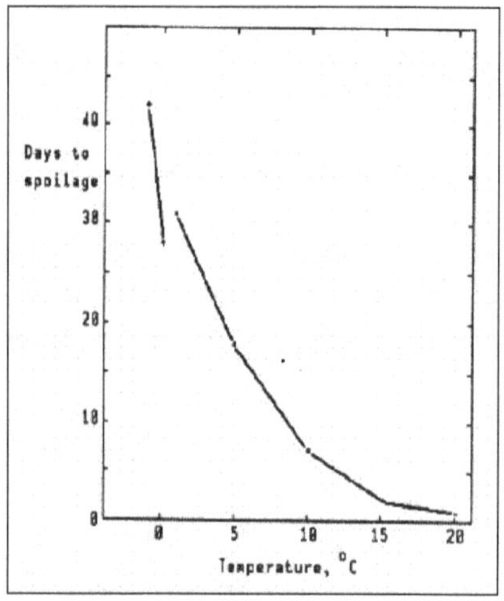

Effect of temperature on time for chicken meat to spoil.

Within the growth range, the rate of growth increases rapidly as the temperature is raised. Conversely, microbial growth rates decrease rapidly as the temperature is lowered and, hence, food spoilage occurs much more slowly. This effect is especially marked near the freezing point. Note in Figure that a drop from about 41 °*F* to about 32 °F will more than double the shelf life (time before spoilage).

Water Activity

Water activity (a_w) is a term describing the availability of water to microorganisms. It is only roughly related to percent moisture. Pure water has an a_w of 1.00, and the atmosphere above the water in a closed container will have an equilibrium relative humidity (ERH) of 100%. If we add an ounce of rocks to a quart of water in such a container, the ERH and the a_w will not change. But if we add an ounce of salt, the ERH will fall to about 98 % and the a_w to 0.98. Rocks do not dissolve in water but salt does, thereby reducing the proportion of water that can enter the atmosphere. Likewise, the amount of water available to microorganisms present in the solution is reduced. Yet the percent moisture is the same in the container with rocks as it is in the container with, salt, namely, 98%.

The GMP regulations for low-acid canned foods defined water activity as the vapor pressure of the food product divided by the vapor pressure of pure water under identical conditions of pressure and temperature. The regulations define low-acid foods as foods, other than beverages, with a finished equilibrium pH value greater than 4.6 and a water activity greater than 0.85.

Table: The water activity (a_w) limits for growth of principal foodborne disease organisms.

Microorganism	Minimal a_w for growth
Salmonella	0.945
Clostridium botulinum	0.95
Clostridium perfringens	0.93
Staphylococcus aureus	0.86
Vibrio parahaemolyticus	0.94

Most bacteria fail to grow in a food or other medium where the a_w is lower than 0.94. Bacteria require a higher a_w than yeasts, which in turn require a higher a_w than molds. Thus, any condition that lowers the a_w first inhibits bacteria, then yeasts, and finally molds. But each species has its limits which are interrelated with other growth factors. Table below gives the a_w limits for growth of principal foodborne disease organisms held under otherwise optimal conditions.

Certain molds and bacteria can grow on fish immersed in saturated salt solution where the a_w is about 0.75. Some molds can grow in foods with a_w 0.62 – 0.65. At these lower limits growth is very slow. The a_w of fully dried foods, such as crackers or sugar, is about 0.10 and such products are microbiologically stable because of this factor alone. The stability of intermediate moisture foods (a_w 0.75 – 0.90), such as dried fruits, jams, and soft moist pet foods, depends on combinations of factors, such as low a_w, low pH, pasteurization, chemical additives, and impervious packaging.

pH

pH is a term used to describe the acidity or alkalinity of a solution. At pH 7, there is an equal amount of acid (hydrogen ion: H+) and alkali (hydroxyl ion: OH-), so the solution is "neutral". pH values below 7 are acidic, while those above 7 are alkaline. pH expresses the H+ concentration logarithmically, that is, in multiples of 10. For example, at pH 5 there are 10 times as many H+ as at pH 6; at pH 3 there are 100 times as many H+ as at pH 5, and so on.

pH has a profound effect on the growth of microorganisms. Most bacteria grow best at about pH 7 and grow poorly or not at all below pH 4. Yeasts and molds, therefore, predominate in low pH foods where bacteria cannot compete. The lactic acid bacteria are exceptions; they can grow in high acid foods and actually produce acid to give us sour milk, pickles, fermented meats, and similar products. Some strains, called Leuconostoc contribute off-flavors to orange juice. The pH values of certain foods are given in table.

Table: Mean pH Values of Selected Foods.

pH Value	Selected Foods
2.3	Lemon juice (2.3), Cranberry sauce (2.3)
3.0	Rhubarb (3.1) Applesauce (3.4), Cherries, RSP (3.4) Berries (3.0 – 3.9), Sauerkraut (3.5), Peaches (3.7), Orange juice (3.7) Apricots (3.8)
4.0	Cabbage, red (4.2), Pears (4.2) Tomatoes (4.3)
4.6	Ravioli (4.6) Pimientos (4.7)
5.0	Spaghetti in tomato sauce (4.9) Figs (5.0) Onions (5.2) Carroes (5.2) Green Beans (5.3), Beans with pork (5.3) Asparagus (5.5), Potatoes (5.5)
6.0	Lima beans (5.9), Tuna (5.9), Tamales (5.9) Codfish (6.0), Sardines (6.0), Beef (6.0) Pork (6.1), Evaporated milk (6.1) Frankfurters (6.2), Chicken (6.2) Corn (6.3) Salmon (6.4)
7.0	Crabmeat (6.8), Milk (6.8) Ripe olives (6.9) Hominy (7.0)

The lowest pH limits for growth of foodborne disease organisms are shown in Table. Many of the investigators who reported these values also determined that adverse factors, such as low temperature or low water activity, increased the minimal pH for growth. But the processor can be sure that these minimal values will prevent growth of these pathogens under any and all circumstances.

Table: The minimal pH minimal for growth of principal foodborne disease organisms.

Microorganism	Growth reported at but not below
Staphylococcus aureus	pH 4.5
Salmonella	4.0
Clostridium botulinum	
Types A and B	4.8
Type E	5.0
Clostridium perfringens	5.0
Vibrio parahaemolyticus	4.8
Bacillus cereus	4.9

Population

A high initial bacterial load increases the likelihood that spoilage will occur under marginal circumstances. This fact is of major importance to the processor of refrigerated foods, the shelf-life of which is enhanced by good sanitation. A high level of spores also increases the possibility that a few will survive to spoil heat processed products.

Oxygen

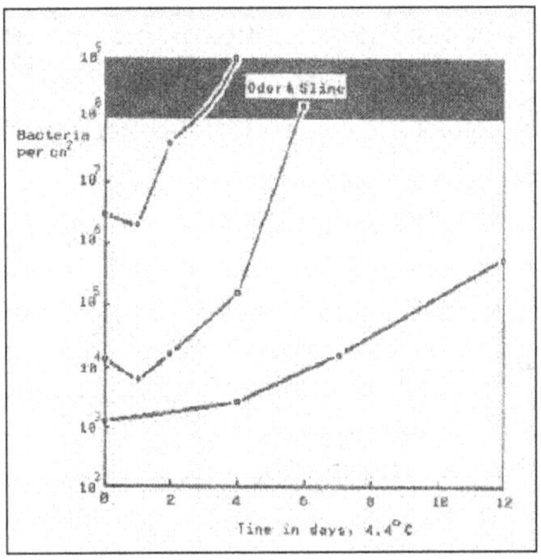	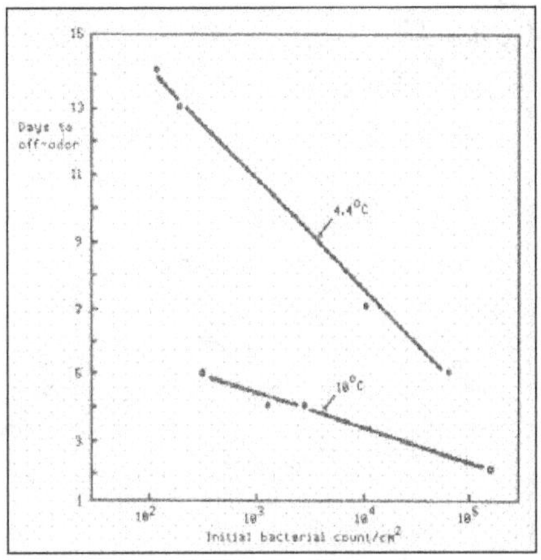
Effect of number of contaminating bacteria on the spoilage of chicken meat.	Effect of numbers of contaminating bacteria on the spoilage time of chicken meat.

Oxygen is essential for growth of some microorganisms; these are called aerobes. Others cannot grow in its presence and are called anaerobes. Still others can grow either with or without oxygen and are called microaerophilic. Strict aerobes grow only on food surfaces and cannot grow in foods stored in cans or in other evacuated, hermetically sealed containers. Anaerobes grow only beneath the surface of foods or inside containers. Aerobic growth is faster than anaerobic. Therefore, in products where both conditions exist, such as in fresh meat, the surface growth is promptly evident, whereas subsurface growth is not.

Lethal Effects of Temperature

Heat is the most practical and effective means to destroy microorganisms. Microbial cell reduction occurs slowly just above maximal growth temperatures. However, the rate of death increases markedly as the temperature is raised. Pasteurization, the destruction of vegetative cells of disease-producing microorganisms, consists of a temperature of 140 °F for 30 minutes, or about 161 °F for 16 seconds. Yeasts, molds, and the vegetative cells of spoilage bacteria also die at pasteurization temperatures. To render log-acid foods commercially sterile requires a retort capable of operating at temperatures above 212 °F. Canners process certain canned foods at 240 °F or 250 °F for a considerable length of time, sometimes an hour or more depending upon the product and can size. Commercial sterility is the destruction and/or inhibition of the organisms of public health significance as well as organisms of non-health significance which could spoil the product. Microbiologists sterilize media at 250 °F (121C) for 15 or 20 minutes. These examples illustrate the need for high temperatures and sufficient time to kill a population of bacteria.

In thermal destruction studies, also called thermal death time studies, the logarithm of the numbers of survivors is plotted against the length of time test cultures are subjected to a given temperature. The result is usually a straight line, although there are many exceptions. The slope of this line becomes steeper as the temperature is increased; indicating that less time is required to kill a population at higher temperatures. It also takes longer to kill a high population of organisms than it does to kill a low population.

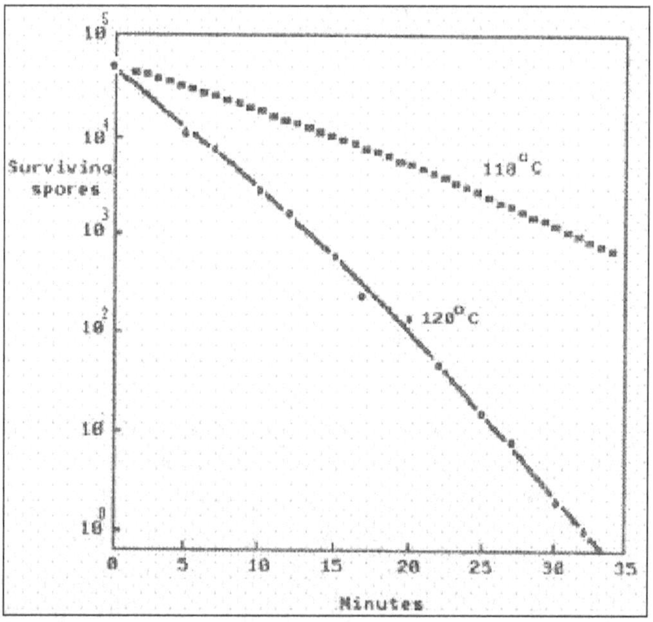

Reduction in numbers of bacterial spores during exposure to lethal temperatures.

The rate of thermal destruction is greater in foods with high a_w than in those with low a_w. Microbial contaminants in dry foods, such as chocolate or dried bone meal, are hard to destroy with heat. The recommended pasteurization process to destroy Salmonella in liquid egg albumen prior to freezing is 140 °F (6 °C) for 3.5 minutes, whereas that for dried egg albumen is 140 (60C) to 158°F (70C) for several days. Riemann was able to kill Salmonella in meat and bone meal more readily at 194° F (90C) after water was added to bring the a_w to 0.90.

Table: The effect of the size of the initial spore population on destruction time.

Microorganism	Spores (number)	Temperature °F (°C)	Destruction time (minutes)
Flat sour #26	45,000	239 (115.9)	62 to 65
	400	239 (115.9)	25 to 28

Clostridium botulinum #90	90,000	221 (105.8)	18 to 20
	900	221 (105.8)	12 to 14

Clostridium botulinum spores are highly resistant to thermal destruction at water activities between 0.2 and 0.4 (dry heat) and are much less resistant to heat at water activities above this range. This finding may be practical for high temperature-short time dry heat sterilization. Other factors that affect the thermal destruction rate of bacteria are the presence or absence of organic matter, oil or fat, pH, strain of organisms, quality of available nutrients, and age of the culture. In general, bacteria are killed more rapidly at lower and higher pH values than in more neutral ranges. In the processing of many foods, careful control of pH is an important factor.

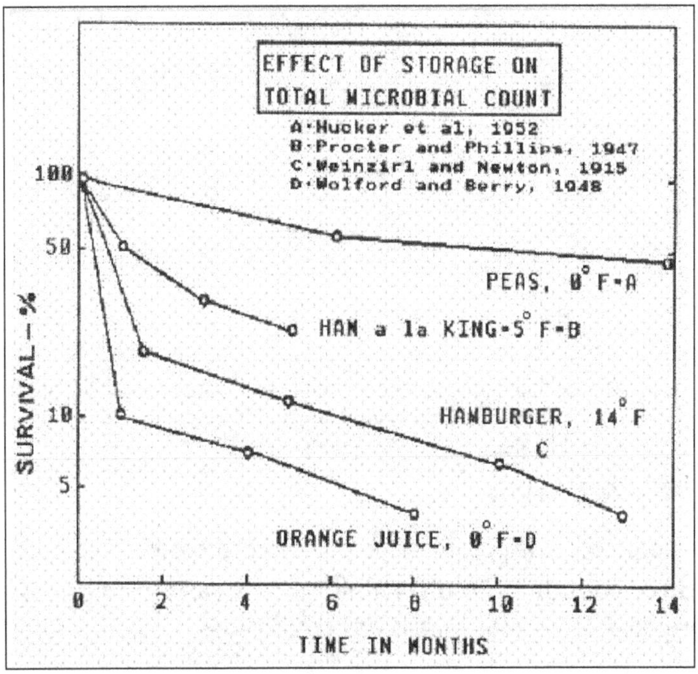

Effect of frozen storage on bacterial level in various foods.

Chilling to temperatures below the growth range, but above freezing, stops reproduction but kills few cells except for extremely sensitive organisms, such as vegetative cells of Clostridium perfringens. Freeze kills part of a microbial population within a few hours and storage continues to be lethal at a much slower rate. The rate of population reduction varies with the nature of the food, the most rapid drop in aerobic plate count ("total count") occurred in orange juice, which is an acid product. Bacterial spores die very slowly, if at all, during freezing and frozen storage. For example, the vegetative cells of Clostridium perfringens generally all die, but the spores survive. Staphylococcus aureus and related organisms survive well, but in most cases there is wide variation of susceptibility among microorganisms, even among closely related species. In any case, freezing is not a dependable means to destroy microorganisms since some cells of the original population almost always survive.

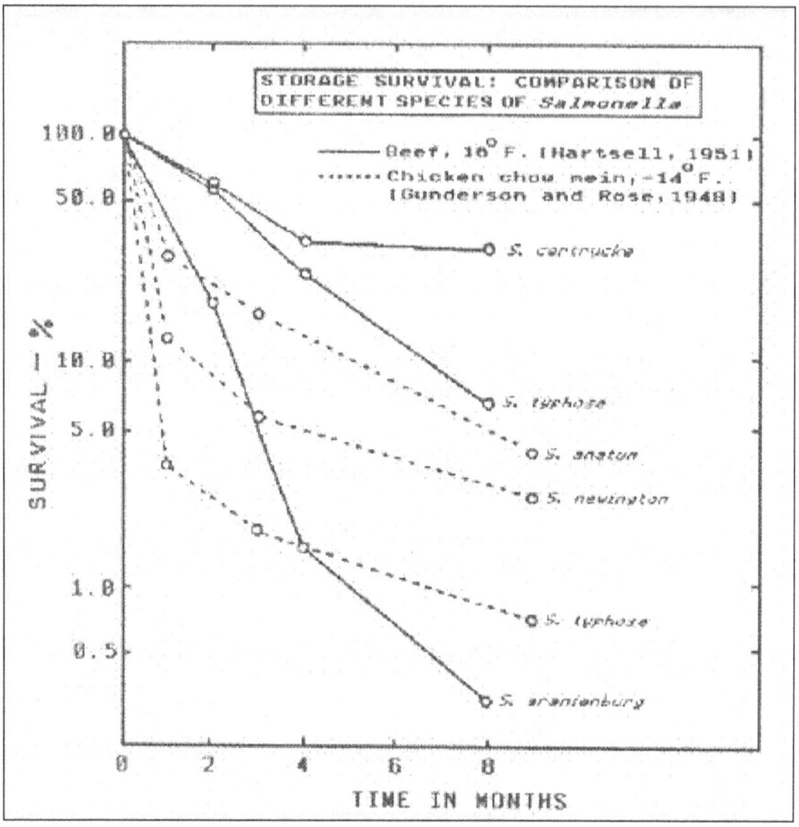

Survival of Salmonella in frozen storage.

The “Indicator” Organisms

The “indicator” organisms are so called because their presence in large numbers in food signifies one of three contamination possibilities: disease bacteria or filth; spoilage or low quality; or preparation under insanitary conditions.

Aerobic Plate Count

The aerobic plate count (APC) measures only that fraction of the bacterial flora that is able to grow to visible colonies under the arbitrary test conditions provided in the time period allowed. It does not measure the total bacterial population in a food sample, but is the best estimate. Altering conditions, such as composition of the agar medium or temperature of incubation, changes the spectrum of organisms that will grow. It is necessary to adhere rigidly to the standardized test conditions that have encouraged some to call the APC a “standard plate count.”

Depending on the circumstances, a high APC may indicate that a food has been grossly mishandled or that it contains a poor quality ingredient. Interpretation depends on knowing what the normal APC is for this food. An abnormal APC indicates that something is out of control. The microbiologist can frequently suggest that cause, thereby aiding the sanitarian. Some of the problems that investigation of a high APC might reveal include:

- Failure of sorting, trimming, washing, and destroying operations to remove or destroy bacteria from raw ingredients adequately.
- Inadequate heat processing.
- Insanitary equipment, particularly near the end of the process.
- The food has reached or is approaching the end of its refrigerated shelf-life.
- The food has been stored at or above room temperature for too long.
- The food is at least partly decomposed.

Coliform Bacteria

The coliform bacteria are non-spore forming rods that occur in large numbers in human and animal feces. They are normally present on raw animal products, such as meats, milk, and eggs, and also occur naturally in soil, water, and surfaces of plants. They are heat sensitive and die rapidly during blanching or pasteurizing. Large numbers of coliforms after a heat process indicate an unacceptable degree of post-heating contamination or indicate time-temperature abuse of the food sufficient to permit growth. High coliform levels warrant investigations to determine the source of contamination or temperature mishandling.

The presence of Escherichia colia, member of the coliform group, in food usually indicates direct or indirect human or animal fecal contamination. Although this may be true in a broad sense, one must not assume a quantitative relationship between the numbers of E. coli and the degree of contamination with feces. E. coli grows well outside the animal body and thrives in unclean food handling equipment.

Food Poisoning

Human illnesses caused by foodborne microorganisms are popularly referred to as food poisoning. The common use of a single classification is due primarily to similarities of symptoms of various food-related diseases. Apart from illness due to food allergy or food sensitivity, foodborne illness may be divided into two major classes, food infection and food intoxication. Food infection results when foods contaminated with pathogenic, invasive, food poisoning bacteria are eaten. These bacteria then proliferate in the human body and eventually cause illness. Food intoxication follows the ingestion of preformed toxic substances which accumulate during the growth of certain bacterial types in foods.

The period of time between the consumption of contaminated foods and the appearance of illness is called the incubation period. The incubation period can range anywhere from less than one hour to more than three days, depending on the causative organisms or the toxic product.

Table: Characteristics of the important bacterial food intoxications and foodborne infections.

Disease	Etiologic Agent	Incubation Period	Symptoms
Botulism	Clostridium botulinum A.B.E.F toxin	Usually 1 to 2 days; range 12 hours to more than 1 week	Difficulty in swelling, double vision, difficulty in speech. Occasionally nausea, vomiting, and diarrhea in early stages. Constipation and subnormal temperature. Respiration becomes difficult, often followed by death from paralysis of muscles of respiration.
Staphylococcal food poisoning	Staphyloccal enterotoxin	1 to 6 hours; average 3 hours	Nausea, vomiting, abdominal cramps, diarrhea, and acute prostration. Temperature subnormal during acute attack may be elevated later. Rapid recovery-usually within 1 day.
Salmonellosis	Specific infection by Salmonella spp.	Average about 18 hours; range 7 to 72 hours	Abdominal pains, diarrhea, chills, fever, frequent vomiting, prostration. Duration of illness: 1 day to 1 week.
Shigellosis (bacillary dysentery)	Shigella sonnei, s. flexneri, s. dysenteriae, s. boydii	Usually 24 to 48 hours; range 7 to 48 hours	Abdominal cramps, fever, chills, diarrhea, watery stool (frequently containing blood, mucus, or pus), spasm, headache, nausea, dehydration, prostration. Duration: a few days.
Enteropathogen zic Escherichia coli infection	Escherichia colisero types associated with infant and adult infections	Usually 10 to 12 hours; range 5 to 48 hours	Headache, malaise, fever, chills, diarrhea, vomiting, abdominal pain. Duration: a few days.
Clostridium perfringens food poisoning	Clostridium perfringens	Usually 10 to 12 hours; range 8 to 22 hours	Abdominal cramps and diarrhea, nausea, and malaise, vomiting very rare. Meat and poultry products usually involved. Rapid Recovery.
Bacillus cereus food poisoning	Bacillus cereus	Usually about 12 hours; range about 8 to 16 hours	Similar to Clostridium perfringens poisoning

Vibrio Parahaemolyticus food poisoning	Vibrio Parahaemolyticus	Usually 12 to 14 hours; range 2 to 48 hours	Abdominal pain, server watery diarrhea, usually nausea and vomiting, mild fever, chills and headache. Duration: 2 to 5 days.

Foodborne Disease Organisms

Escherichia coli

A few of the E. Coli strains found in human feces are in themselves pathogenic, causing infection and disease. These are called Enteropathogenic E. Coli or EEC. In one extensive study of the feces of food handlers, 6.4% of the workers harbored the EEC organisms as carriers.

Staphylococcus Aureus

S. aureus, commonly referred to as "staph," is normally present on the skin, the mucous membranes, and in pimples and boils of human beings and other animals. It is nearly always present in small numbers in raw meats and in foods handled extensively by human hands. The food poisoning strains generally come from human sources. Pasteurizing or cooking destroys the organism, but not its toxin. Foods contaminated by staph organisms can cause food poisoning after the organisms have been destroyed by heat. The presence of staph in a cooked food has two levels of significance:

- Low numbers (not over a few hundred per gram) indicate the degree of contact with human skin or nasal mucous, cross-contamination from raw meat, or survivors of a larger population.
- High numbers (100,000 or more per gram) indicate that the bacteria were allowed to grow in the food, thereby creating the potential serious hazard of the presence of toxin.

Keeping foods completely free from staph contamination is often difficult or impossible. Therefore, the processor should store the food at temperatures that preclude the growth of staph. It is only during growth that staph forms the toxin. An epidemiological investigation to determine the source of the organism is tedious, but visual inspection of workers' hands can be useful. The well-informed sanitarian will also seek time-temperature abuses of foods contaminated with staph. The National Academy of Science's National Research Council has listed the following steps to limit the incidence and level of staph in foods:

- Reduce direct and indirect exposure of foods, particularly cooked foods, to human contact as much as possible. If handling is necessary, use sanitary rubber or plastic gloves, or sanitize hands. Persons with infected cuts, abrasions, boils, or pimples should never handle cooked foods.

- Test raw materials and eliminate production lots that contain high levels of S. aureus.
- Process to destroy the microorganisms.
- Eliminate cross-contamination from raw to cooked food.
- Keep cooked foods no longer than 2 to 3 hours between 40 °F and 140 °F.

Control of staph growth in fermented foods, such as cheese or sausages, requires controlling a number of processing factors. Low pH, relatively high levels of lactic bacteria, salt, and nitrite help to inhibit toxin formation.

Salmonella

Salmonella infection, or salmonellosis, is almost always caused by eating contaminated food or drink. Contamination originates from the intestinal tract of human beings or animals that harbor Salmonella organisms. Most adults can resist infection from a few cells, but become ill when ingesting millions. Infants, the aged, and the infirm are much more sensitive and can be affected by a few Salmonella cells. After recovery, the victim may remain a carrier for a period varying from a week to permanency.

Domestic animals, such as dogs, poultry, swine, horses, sheep, and cattle are carriers of these pathogens. Carriers show no outward symptoms of the disease at the time of slaughter. As long as abattoirs continue to receive Salmonella carriers for slaughter, Salmonella contamination of the finished raw meat is inevitable. Even with apparently satisfactory sanitation, slaughtering and dressing procedures may spread traces of feces from a carrier animal to subsequently slaughtered animals by way of equipment, water, and hand contact.

Salmonella is often discussed as if it were a single organism. There are actually more than 1,300 serotypes identified within the genus Salmonella. All are quite sensitive to heat, so freshly pasteurized or cooked foods are free of the organism. The principal routes of its entry into cooked foods are cross-contamination from raw foods or animals (via hands, equipment, air, water), recontamination from human carriers, or gross undercooking. Regulatory agencies are quick to institute seizures, recalls, and other legal action against products and firms shipping Salmonella-contaminated processed foods.

Dry and semi-dry fermented sausages rarely cause food borne diseases. However, recent investigations by USDA have shown that Salmonella can survive the fermentation and drying process. Salmonella in natural animal casings likewise survives short periods of salting, but dies more rapidly in acidified or alkalized casings.

Salmonella can also grow outside the animal body when conditions are favorable. For this reason, it has appeared in a wide variety of foods and feeds, in addition to meat and poultry products. Some of these are brewer's yeast, coconut meat, cochineal dye, dried or frozen eggs, noodles, custards, dried animal feeds, cottonseed flour, candy, chocolate, dried milk, fish and shellfish, cream-filled pastries, sausage casings, and watermelon. The NAS-NRC has made extensive recommendations for evaluation, control, and eradication of the Salmonella problem.

Costridium Botulinum

C. botulinum produces a rare but often fatal disease called botulism. It is caused by a neurotoxin produced during growth in the absence of air. Except in the case of infantile botulism the intact spores are harmless. Infants ingesting spores, usually from honey, have developed symptoms of botulism. Botulism usually occurs after a food containing the preformed toxin has been eaten, but sometimes the organism infects wounds, forming the toxin in the muscle of the victim. There are seven types of C. botulinum (A to G), of which four (A, and B associated with meats and vegetables, E, marine environment and F) cause human disease. Only once has type C been reported to cause human illness. Type G is a new incompletely studied discovery.

Fortunately, the toxins, regardless of type, have very little resistance to heat and are inactivated by boiling for 10 minutes. Thus, all freshly, but adequately, cooked foods are safe. All C. botulinum strains can form spores which exhibit varying resistance to heat. The spores of types A and B are highly resistant. Spores of type E die in a fraction of a minute at 212 °F. The canning industry, under the technical leadership of the National Food Processors Association (formerly the National Canners Association), has established times and temperatures of retorting necessary to insure the commercial sterility of low-acid canned foods. The NFPA also submitted to the FDA the initial petition which eventually developed in the GMP regulations for low-acid canned foods.

Botulinum spores are widely distributed in soils. Type A predominates in the western states and in New England; type B, in the eastern and southern states. Type E is usually associated with marine or fresh water environments throughout the world and is psychrotropic. Type F has been isolated too rarely to establish its distribution pattern.

C. botulinum will not grow below pH 4.8. Therefore, botulism is a concern only in low acid foods, which are defined as foods with a finished equilibrium pH greater than 4.6. The majority of outbreaks occur from home canned vegetables, meats, fish, and over-ripe fruits. Canned cured meats contain salt and nitrite. The preservatives protect against the outgrowth of botulinum spores that may have survived the minimal processing, which is frequently at or below boiling.

There have been 34 outbreaks of type E botulism among fish products prepared in the U.S. and Canada. Most have been smoked or lightly salted products. The FDA isolated botulinum types B, E, and F from pasteurized meat of the blue crab. The NAS-NRC has reviewed steps to minimize the possibility of out-breaks from smoked fish and FDA has published regulations designed to control the problem.

Clostridium Perfringens

C. perfringens is a spore-forming organism which, like botulinum, grows only in the absence of air. It grows best in meat or poultry dishes, stews, or gravies kept warm. Such foods meet its exacting nutritional requirements and the warm holding temperature, up to 122 °F, encourages its growth. The spores themselves are harmless, but the vegetative cells, which can grow to enormous numbers in these foods, form spores in the intestinal tract of the victim. During the sporulation process, the remainder of the vegetative cell dissolves, releasing the poison that causes illness. The vegetative cells which cause the disease are very delicate. They can be destroyed or reduced to low, safe levels by cooking or freezing. The spores are widely distributed in nature and are present in small numbers in various foods. They occur in feces, soils, dust, water, marine sediments, raw foods, and even cooked foods.

C. perfringens poisoning is a problem specific to the food service industry. Only proper temperature control prevents the problem. A good rule of thumb is to keep ready-to-eat moist foods below 40 °F or above 140 °F. Time-temperature abuse is a severe health hazard. Since the spores are everywhere, epidemiologic investigation of strains to determine the source of spores is a relatively futile exercise. However, if serological tests show that the same types are present in the victim's food and feces, a particular dish can be incriminated. Unfortunately, the biological materials (antisera) for this purpose are not yet commercially available. Therefore, the determination that large numbers of C. perfringens cells are present remains the most suitable investigative test.

Bacillus Cereus

B. cereus is a spore-forming organism that grows in the presence of oxygen and is widely distributed in most raw foods. Since the spores survive boiling for several minutes, they remain viable in cooked foods in small numbers. The organism does not compete well with other bacteria in raw foods, but in moist, cooked dishes held warm (up to 122 °F), it grows to millions per gram in a few hours. Under these conditions the food becomes poisonous. B. cereus grows well in a wide variety of cooked foods, such as meats, poultry, sauces, puddings, soups, rice, potatoes, and vegetables. The disease is similar to that of perfringens, although the mechanism of the disease is unknown. Adults have rather mild symptoms, but small children may become seriously ill. In most instances, the victims recover quickly and do not seek medical attention. Therefore, only large outbreaks are reported and become part of the statistical record. Similar to C. perfringens, B. cereus is primarily a concern of the food service industries. The appropriate control is to keep hot foods hot (over 140 °F) and cold foods cold (under 40 °F). Epidemiologic investigation of strains to determine the source of the spores proves equally futile.

Vibrio Parahaemolyticus

V. parahaemolyticus is a non-spore forming, slightly curved rod, closely related to the organism that causes cholera. It is widely distributed and grows in brackfish waters, estuarine sediments, raw fish, and shellfish throughout the world. It competes well with spoilage organisms at temperatures of 41 °F or above. It occurs in greatest numbers in the summer when higher temperatures engender rapid growth.

V. parahaemolyticus is the principal cause of food poisoning in Japan where raw fish is regularly consumed. Elsewhere, the disease occurs less frequently because the organism dies readily during pasteurization or cooking. Nevertheless, cooked seafoods can be recontaminated from water or raw seafood. The first confirmed outbreaks in the United States occurred in 1971 and 1972 from crabmeat, shrimp, and lobster. In one Japanese outbreak, 22 people died and 250 others became ill.

The human pathogenicity of the organism is determined by culturing it on a special medium, a salt agar containing human blood. If the organism can grow and destroy blood cells on this medium, the so-called Kanagawa test, it is labeled "Kanagawa positive" and assumed capable of causing human disease. The Japanese have found that about 1% of the strains of V. parahaemolyticus from waters near their shores are Kanagawa positive. On the other hand, reported that up to 90% of the strains from U.S. estuarial waters are Kanagawa positive. However, the significance of the Kanagawa test is not fully understood. To reduce the incidence of these outbreaks, the seafood industry should:

- Hold raw seafoods at or below 40 °F.
- Keep cooked seafoods carefully apart from raw seafood, sea water, insanitary equipment, and unclean containers.
- Hold cooked seafood below 40 °F or above 140 °F.

Listeria

Before the 1980's most problems associated with diseases caused by Listeria were related to cattle or sheep. This changed with food related outbreaks in Nova Scotia, Massachusetts, California and Texas. As a result of its widespread distribution in the environment, its ability to survive long periods of time under adverse conditions, and its ability to grow at refrigeration temperatures, Listeria is now recognized as an important food borne pathogen.

Immunocompromised humans such as pregnant women or the elderly are highly susceptible to virulent Listeria. Listeria monocytogenes is the most consistently pathogenic species causing listeriosis. In humans, ingestion of the bacteria may be marked by a flu-like illness or symptoms may be so mild that they go unnoticed. A carrier state can develop. Following invasion of macrophages virulent strains of Listeria may then multiply, resulting in disruption of these cells and septicimia. At this time the organism has access to all parts of the body. Death is rare in healthy adults; however, the mortality rate may approximate 30% in the immunecompromised, new born or very young.

Listeria monocytogenes is a special problem since it can survive adverse conditions. It can grow in a pH range of 5.0-9.5, in good growth medium. The organism has survived the pH 5 environment of cottage cheese and ripening Cheddar. It is salt tolerant surviving concentrations as high as 30.5% for 100 days at 39.2 °F. But only 5 days if held at 98.6 °F. The key point is that refrigeration temperatures do not stop growth of Listeria. It is capable of doubling in numbers every 1.5 days at 39.2 °F. Since high heat, greater than 175 °F, will inactivate the Listeria organisms, post-process contamination from environmental sources then becomes a critical control point for many foods.

Yersinia Enterocolitica

Even though Yersinia enterocolitica is not a frequent cause of human infection in the U.S., it is often involved in illness with very severe symptoms. Yersiniosis, infection caused by this microorganism, occurs most commonly in the form of gastroenteritis. Children are most severely affected. Symptoms of pseudo-appendicitis have resulted, in many unnecessary appendectomies. Death is rare and recovery is generally complete in 1 – 2 days. Arthritis has been identified as an infrequent but significant sequela of this infection.

Y. enterocolitica is commonly present in foods but with the exception of pork, most isolates do not cause disease. Like Listeria this organism is also one that can grow at refrigeration temperatures. It is sensitive to heat (122 F., sodium chloride (5%) and acidity (pH 4.6), and will normally be inactivated by environmental conditions that will kill salmonellae.

Campylobacter Jejuni

C. jejuni was first isolated from human diarrheal stools in 1971. Since, then it has continually gained recognition as a disease causing organism in humans. C. jejuni enteritis is primarily transferred from animal origin foods to humans in developed countries. However, fecal contamination of food and water and contact with sick people or animals predominates in developing countries. Although milk has been most frequently identified throughout the world to be a vehicle for Campylobacter, one anticipates that future investigations will identify poultry and its products and meats (beef, pork and lamb) as major reservoirs and vehicles. C. jejuni dies off rapidly at ambient temperature and atmosphere, and grows poorly in food.

The principles of animal science will play a significant role in the control of this ubiquitous organism. Hygienic slaughter and processing procedures will preclude cross-contamination while adequate cooling and aeration will cause a decrease in the microbial load. In addition, thorough cooking of meat and poultry products followed by proper storage should assist in maintaining food integrity and less contamination.

Mycotoxins

Mycotoxins are harmful byproducts from molds that grow on foods and feeds. They have caused severe illness and death in animals for centuries. They first came to the attention of modern scientists in 1960 when 100,000 turkey poults died in England after eating moldy peanut meal from Africa and South America. The mycotoxins involved were later shown to be aflatoxins, a group of closely related organic compounds that can cause acute disease and death. Stimulated by these first discoveries and by research in antibiotics, investigators have discovered dozens of mold strains which produce a wide variety of mycotoxins that affect animals. There are now about 60 identified toxins. Of these, only a few have been designated human food contaminants. These numbers will likely increase as mycotoxin investigations continue and identification methods are improved.

Historically, mycotoxins have been associated with human poisoning and even death. Ergot is among the first mycotoxins recognized as affecting human beings. It is produced by a mold growing on cereal grains. Ergot poisoning occurred in the Rhine Valley in the year 857 and has been reported several times since. The most recent outbreak was in 1951 in southern France. Many Russians died during World War II from eating moldy grains. The Japanese have reported human toxicity from eating moldy rice; the disease caused severe liver damage, hemorrhaging, and some fatalities.

Although such incidents are rare occurrences, there is evidence that low dietary levels of aflatoxins contribute to cancer of the liver in human beings. Extensive laboratory studies have also shown that even at very low dietary levels, aflatoxin can produce liver cancer in rats, mice, monkey, ducks, ferrets, and rainbow trout. Epidemiological studies in Southeast Asia and Africa have related a high incidence of human liver cancer to aflatoxin levels up to 300 parts per billion (ppb) in 20% of the food staples, and 3 to 4 ppb in 7% of the foods as eaten. In one geographical area, 95% of the corn and 80% of the peanuts contained aflatoxin at an average level of 100 ppb.

Although there is no direct evidence that aflatoxins cause human liver cancer in the United States, FDA is concerned about the effect of long-term, low-level consumption of a known, highly carcinogenic substance in our food supply. FDA established an informal defect action level tolerance of 30 ppb on peanuts and peanut products in 1965. With improved harvesting, storage, and sorting practices developed by USDA and industry, the level of aflatoxins contamination gradually declined and FDA lowered the informal action level to 20 ppb in 1969. FDA proposed in the Federal Register of December 6, 1974, a regulation establishing a tolerance of 15 ppb for total aflatoxins in shelled peanuts and peanut products used as human food. Today the limits are 0.5 ppb for milk, 20 ppb for food, and 100 ppb for feed.

Molds which form mycotoxins can be present on any food not heated in a closed container. One must assume, therefore, that they are present and capable of producing, toxin if conditions permit. But finding a toxigenic mold in a food does not imply that the food contains a mycotoxin. Conversely, the absence of visible growth of an aflatoxin producing mold does not mean toxin is absent since aflatoxins may be produced when there is little visible mold growth.

There are several ways to determine whether molds growing in an abused food will produce mycotoxins. The food can be held with its naturally contaminating molds, or inoculated with a toxigenic strain, and kept until the molds develop. The food can then be tested for the presence or absence of toxin. Such experiments have demonstrated that molds produce mycotoxins on a large variety of cereal grains and seeds, dry beans and fruits, spices, nuts, and cured meats. As do bacteria, molds have moisture, temperature, and nutritional requirements for optimal growth and toxin production. In most cases the initial mold invasion occurs in the fields before or during harvest. Mold growth continues during storage if the moisture content and storage temperatures remain high.

Aflatoxin has been found throughout the world on corn, barley, copra, cassava, spices, dry milk, tree nuts, cottonseed, peanuts, rice, wheat, and grain sorghum. In the U.S. it has been found in corn, figs, grain sorghum, cottonseed, peanuts, and certain tree nuts. The industry has relied on electronic and visual sorting methods, as well as blowing and vacuuming, to control aflatoxin levels in walnuts arid pecans. Corn mill operators use a high intensity ultraviolet ("black") light to detect possible aflatoxin contamination. Roasting reduces the level of aflatoxin up to 50% in some cases.

The universal solution to the problem is eliminating conditions that permit mold growth, whenever it is feasible to do so, and thereby preventing the formation of mycotoxins. In some cases (corn, peanuts) mold growth and toxin production occur before harvest. Insect and bird damaged corn kernels are very susceptible; therefore, controlling these pests will help alleviate mold problems. For most susceptible foods, the critical period is immediately following harvest, during storage and initial drying when the moisture content is high enough to allow mold growth.

Microbiological Spoilage of Foods

Bacterial Spoilage of Foods

Although many bacteria can grow in foods, only some specific groups are responsible for their spoilage. The survival, growth, and occurrence of these microorganisms in foods are affected by many different factors, including storage temperature, oxygen availability, food composition, pH, thermal treatment, and competing microbiota, among others.

Table: Main bacteria involved in the spoilage of specific food groups.

Types of food	Main groups responsible for spoilage
Refrigerated, fresh meats (aerobic atmosphere)	Pseudomonas spp. and other psychrotrophs.
Refrigerated, fresh meats (vacuum or modified atmosphere)	Brochothrix thermosphacta, Clostridium spp., Shewanella putrefaciens, Enterobacteriaceae, and lactic acid bacteria (LAB).
Meat products (cured, salted, or fermented)	Micrococcus, Staphylococcus, Enterococcus, and LAB
Seafood (fresh)	S. putrefaciens, Aeromonas spp., Photobacterium phosphoreum, and Pseudomonas spp.
Seafood (processed and salted)	Brochothrix thermosphacta, LAB, Enterobacteriaceae, and Vibrio spp.
Refrigerated, raw milk	Pseudomonas spp., Bacillus, Micrococcus, Aerococcus, Lactococcus, and Enterobacteriaceae.
Ultra-high-temperature milk	Bacillus spp.
Cheeses	Psychrotrophs, Coliforms, Clostridium tyrobutyricum, Clostridium sporogenes, Lactobacillus, and Leuconostoc.
Butter and cream	Pseudomonas spp. and Coliforms.
Eggs and related products	Serratia, Pseudomonas, Proteus, and Aerobacter
Vegetables	Erwinia carotovora, Pseudomonas spp., Corynebacterium, Xanthomonas campestris, and LAB.
Fruits	Pseudomonas spp., Erwinia, Lactobacillus spp., Xanthomonas, and Acidovorax.
Fruit juices	Alicyclobacillus spp. and Propionibacterium sp.
Bakery products	Bacillus spp.
Canned foods	Thermoanaerobacterium thermosaccharolyticum, Moorella thermoacetica, Desulfotomaculum nigrificans, Clostridium butyricum, and the facultative anaerobe Geobacillus stearothermophilus.

Bacterial spoilage occurs faster and more evidently in high-protein foods, such as meat, milk, fish, and dairy products. Most of these foods are rich in nutrients and have high water activity as well as pH values close to 7.0. However, the occurrence of certain microorganisms in these foods is mainly promoted by temperature and atmosphere. Temperature is a limiting factor as it can favor the growth of some spoilage microbial groups and impair the growth of others. For example, low temperature favors the growth of psychrotrophic bacteria over the growth of mesophiles and thermophiles. In this context, Pseudomonas spp., a strictly aerobic psychrotrophic bacterium, prevails as the main spoilage microorganism in chilled fresh meat and may cause color, odor, and flavor changes as it degrades amino acids and fats.

They create a slimy layer on the surface of the food as they grow. This species, together with S. putrefaciens, Photobacterium phosphoreum, and some Vibrio spp., is the main microorganism involved in the spoilage of seafood. Moreover, Pseudomonas spp., together with other psychrotrophic species, is responsible for proteolysis and hydrolytic rancidity of raw milk and dairy products and also for changes on the surface (slime), pigmentation, and odor of cheeses. Despite the high fat content of butter and cream, these can also be spoiled by this group of bacteria. In addition to changes promoted by psychrotrophic and LAB, cheese may also be spoiled by coliforms (early blowing) or gas-producing, anaerobic, spore-forming bacteria, such as Clostridium tyrobutyricum (late blowing). Eggs are a very popular food and are also used as ingredients in many foods. They have many barriers that prevent the access and growth of microorganisms within them. The prevailing microbiota in the spoilage of this product is composed of Gram-negative bacteria, including Pseudomonas spp., which rot the product and change its color.

A small change in the way a food is packaged, such as the use of vacuum or modified atmosphere, can change the profile of the spoilage microorganisms from aerobic to facultative or strictly anaerobic. In meats and meat products stored under these conditions, spoilage is caused by LAB, Enterobacteriaceae, and sometimes Brochothrix thermosphacta. Brochothrix thermosphacta is an important species associated with the spoilage of meats and meat products and grows in both aerobic and anaerobic conditions.

Other changes in the spoilage microbiota of high-protein foods may be induced by changing water activity and adding a competing microbiota. For example, curing, salting, and fermenting meat products inhibits the growth of the natural spoilage microbiota and promotes the growth of bacteria better adapted to these conditions, such as species of the genera Micrococcus and Staphylococcus and some LAB.

The presence of complex nutrients in fruits and vegetables, such as cellulose, polysaccharides, hemicellulose, and pectin and in some cases the acidic pH, limits the bacterial spoilage of these foods. However, some bacteria are capable of producing extracellular lytic enzymes that degrade these polymers, releasing nutrients for their growth. This softens the flesh and produces acids, alcohols, and metabolites with unpleasant odors and flavors. The main bacterium species responsible for this deterioration is Erwinia carotovora, present in nearly all vegetables and some fruits. However, the intrinsic factors of bakery products limit bacterial growth and favor the prevalence of filamentous fungi as the main spoilage microorganisms. However, bacteria from the genus Bacillus spp. may cause a type of spoilage in bread known as ropiness, which is characterized by brown and black stains, release of a rotten fruit odor, and a sticky and moist breadcrumb. Bacillus subtilis is the main species responsible for this spoilage, but other members of the genus, such as Bacillus licheniformis and Bacillus megaterium, have also been identified causing this spoilage.

Thermal treatment preserves and insures the safety of some types of food, but it also allows specific spoilage microorganisms to prevail in the treated food. In ultra-high temperature milk, spoilage is caused mainly by spore-forming bacteria of the genus Bacillus, which resist thermal treatment and produce enzymes that promote sweet coagulation, gelification, and odor changes Alicyclobacillus spp. and Propionibacterium cyclohexanicum are the main microorganisms associated with the spoilage of fruit juices that survive thermal treatment and may cause off-flavors, off-odors, discoloration, and turbidity. Canned foods submitted to industrial sterilization methods are spoiled exclusively by facultative or strictly anaerobic spore-forming bacteria, such as G. stearopthermophilus, Clostridium sporogenes, and Desulfotomaculuns nigrificans, among others.

Yeast Spoilage of Foods

Yeasts are very important unicellular eukaryotic microorganisms involved in both food and beverage production and spoilage. These microorganisms spoil foods with high sugar or salt contents, low pH, or other characteristics that give them a competitive advantage over bacteria. Most yeasts produce extracellular enzymes, such as proteases, lipases, amylases, and pectinases, and also volatile and non-volatile metabolites that affect the sensory characteristics of food, especially flavor and texture. The main genera associated with food spoilage are Saccharomyces, Candida, Zygosaccharomyces, Debaryomyces, Rhodotorula, and Pichia.

Table: Main yeasts involved in the spoilage of specific food groups.

Types of food	Main groups responsible for spoilage
Fresh meats	Pseudomonas spp. and other psychrotrophs.
Refrigerated, fresh meats (vacuum or modified atmosphere)	Candida spp., Rhodotorula spp., Debaryomyces spp., Trichosporon spp., and Torulopsis spp.
Dry and salted meat products	Debaryomyces hansenii, Yarrowia lipolytica, Candida, Trichosporon, Cryptococcus, and Rhodotorula.
Yogurts	Saccharomyces cerevisiae and Hansenula anomala.
Cheeses	Candida spp., Kluyveromyces marxianus, Debaryomyces hansenii, and Pichia spp.
Butter	Candida parapsilosis, Candida zeylanoides, and Yarrowia lipolytica.
Fruits and vegetables	Saccharomyces, Candida, Zygosaccharomyces, Torulopsis, Rhodotorula, Hansenula, Debaryomyces, and Pichia spp.
Bakery products	Saccharomyces, Debaryomyces, Kluyveromyces, Pichia, Candida, and Zygosaccharomyces.
Chocolate, honey, and candy	Zygosaccharomyces spp., S. cerevisiae, Torulopsis apicola, Hansenula anomala, and Kloeckera apiculata.
Carbonated soft drinks	Saccharomyces spp., Zygosaccharomyces spp., Torulaspora delbrueckii, Pichia anomola, Dekkera spp., and Candida spp.

Yeasts play a small part in the spoilage of meats. Despite the competition with bacteria, some yeast manage to grow on the surface of fresh meats, whereas others may have their growth promoted by changes in intrinsic and extrinsic factors that inhibit bacterial growth. Yeasts are also known for their important role in the dairy industry, especially for the production of some fermented products, maturation of some cheeses, and whey fermentation for bioactive compound production. However, yeasts are the main cause of yogurt and fermented milk spoilage because the low pH provides a selective environment for their growth. Their presence is associated with off-flavors and gas production. These microorganisms also spoil cheeses, and most contamination stems from the brine used in the production process. Yeasts are normally not involved in butter spoilage, but they have already been detected on spoiled butter.

Yeasts play an important role in the spoilage of fruits and vegetables, especially because of the exposure of these foods to the environment and their minimal processing. Some of their intrinsic factors, such as pH, also limit bacterial growth. Yeasts are also involved in the deterioration of bakery products, causing white and pink stains or fermenting the carbohydrates, with subsequent production of volatile compounds and alcoholic odor. In carbonated beverages, the growth of spoilage yeasts is usually characterized by a great production of gas, packaging distension and rupture, abnormal effervescence, excessive turbidity, sedimentation, off-flavors, and off-odors. Osmophilic yeasts, together with xerophilic fungi, are the only microorganisms involved in the spoilage of high sugar foods with low water activity, such as chocolates, confectionary products, and honey. Zygossacharomyces bailli is involved in spoilage of low-pH beverages containing preservatives and high-sugar foods because it has developed mechanisms to adapt these stress factors. These yeasts play a major role in spoilage of beverages, such as fruit juices, soft drinks, chocolate fillings, confectionary creams, etc.

Filamentous Fungi Spoilage of Foods

Filamentous fungi are capable of growing on many different foods, such as grains, meats, milk, fruits, vegetables, seeds, and high-fat products. They are an important group of food spoilage organisms and cause significant economic losses in agriculture and the food industry. Spoilage caused by filamentous fungi may manifest as discoloration, off-flavors, loss of structure, loss of texture, formation of visible mycelium, and production of volatile compounds, all of which affect the quality of foods and beverages. These obligate aerobic microorganisms are capable of growing in wide ranges of pH, temperature, and water activity and of using a great variety of substrates as food. Filamentous fungi grow more commonly in products with low pH and water activity. They are located mainly on the surface of the food because of their dependence on oxygen. The main food spoilage fungi are Penicillium, Aspergillus, Rhizopus, Mucor, Geotrichum, Fusarium, Alternaria, Cladosporium, Eurotium, and Byssochlamys. Some of these species are also known for their ability to synthesize secondary toxic metabolites called mycotoxins, constituting a problem for agribusiness and the food industry.

Table: Main filamentous fungi involved in the spoilage of specific food groups.

Types of food	Main groups responsible for spoilage
Salted fish and similar products	Wallemia, Hortaea, Aspergillus, Penicillium, Polypaecilum, Eurotium, Basipetospora, Cladosporium, and Scopulariopsis
Cheeses	Penicillium, Cladosporium, Byssochlamys nivea, Talaromyces avellaneus, Neosartorya fischeri var. spinosa, and Eupenicillium brefeldianum
Yogurts	Penicillium and Aspergillus
Meat and meat products	Aspergillus spp., Penicillium spp., Mucor spp., Cladosporium spp., and Eurotium spp
Fruits and vegetables	Penicillium, Phytophthora, Alternaria, Botrytis, Fusarium, Cladosporium, Phoma, Trichoderma, Aspergillus, Alternaria, Rhizopus, Aureobasidium, and Colletotrichum

Controlling Microbiological Spoilage of Foods

Avoiding or retarding spoilage of raw and processed foods but at the same time preserving the sensory and nutritional characteristics of foods is a great challenge for the industry. The methods used for controlling microbial spoilage include preventing access of the organisms to the foods, removing their cells or spores, inhibiting their growth, and using thermal and nonthermal methods to inactivate them.

Table: Methods for controlling the microorganisms involved in food spoilage.

Principle of preservation	Methods
Prevent access of the microorganism to the food	Good agricultural practices Good manufacturing practices Appropriate hygienic and sanitary conditions.
Microorganism removal	Washing of surfaces Centrifugation Filtration by membranes.
Growth inhibition	Cooling or freezing Reduce water activity (drying and addition of solutes) Food acidification Addition of inhibitors (weak organic acids, nitrites, nitrates, sulfites, and bacteriocins) Microbial competition Atmosphere modification (vacuum or modified atmosphere packaging).
Microbial inactivation	Thermal treatments (pasteurization, sterilization, blanching, and thermization) Radiation High hydrostatic pressures Pulsed electric field processing Ultrasound.

The good manufacturing and handling practices are very important means of reducing the baseline microbial load and increasing shelf life. They are considered the minimum hygienic requirements for the production of any type of food. These practices should be used throughout the entire food production chain, from production of raw materials to consumer consumption. Other methods of preventing spoilage include: (1) washing the raw materials before processing, which removes microbial cells from its surface, (2) centrifuging, (3) or using membrane filtration processes, which can only be used on certain food groups. In addition to the above-mentioned methods, foods can also be preserved by manipulating factors that not only influence microorganism growth and survival but are also safe for consumers. Control of microbial growth by changing intrinsic and extrinsic factors, also called barrier theory, is the main way of reducing food spoilage.

The most important barriers used for food preservation are temperature (high and low), low a_w, acidity (pH), redox potential (Eh), preservatives, and other organisms (microbial competition). These microbial control methods use changes in food characteristics and storage conditions to prevent or reduce microbial growth. Combining different sublethal conditions to prevent microbial growth is a successful approach mainly because it does not change the sensory characteristics of the foods significantly. The great disadvantage of these methods is that microorganisms can resume growth if favorable growth conditions occur.

Some food preservation methods focus on microbial inactivation. They can be classified as thermal or non-thermal methods of food preservation. The thermal methods, such as pasteurization and commercial sterilization, are widely used methods to efficiently inactivate pathogens and reduce the load of spoilage microorganisms. Despite this, they can also change the nutritional and sensory properties of the food. To replace these methods and meet the increasing consumer demand for healthier products, many non-thermal methods are being developed and used. The main non-thermal methods are radiation, high hydrostatic pressure, pulsed electric field, and ultrasound. Many of these methods are still difficult to implement because of equipment cost, need of trained personnel, and consumer distrust.

2

Food Technology: Processing and Preservation

The branch of science which deals with the production, processing, preservation, packaging, etc. of food is termed as food technology. This chapter has been carefully written to provide an easy understanding of the varied facets of food processing such as stages and methods of food processing, canning and bottling of foods, etc.

Food technology is the application of food science to the selection, preservation, processing, packaging, distribution, and use of safe food. The field also involves techniques and processes that are used to transform raw materials into food. Extensive research goes behind making food items edible as well as nutritious. The complex food system is changing continuously. Processing of food has developed over the time making food easily accessible, more nutritious, safe and affordable. With the aid of various disciplines in food science and technology, raw materials are being converted to edible foods.

Reduced diseases, improved safety and quality, variety, less cost, less wastage are some of the advancements in the field. But the growing need of food with specific attributes due to population hike is of concern. Biotechnology field has made progress to meet the increasing demand by improving crop production. Maintaining human health and wellness is of much importance. Newer technologies and detecting mechanisms can be anticipated in future with the evolving progress in nanotechnology. Minimizing risks and maximizing the benefits should be the main objective. Experts in Food science and technology should work hand in hand to get the maximum output. The approach should be balanced and persistent to safeguard both nature and human health.

Food Processing

Food processing is any of a variety of operations by which raw foodstuffs are made suitable for consumption, cooking, or storage. A brief treatment of food processing follows. Food processing generally includes the basic preparation of foods, the alteration of a food product into another form (as in making preserves from fruit), and preservation and packaging techniques.

A number of food-processing innovations have even resulted in new products, such as concentrated fruit juices, freeze-dried coffee, and instant foods. Foods and food supplements have also been processed from such hitherto untapped sources as oilseeds (chiefly protein-rich soybeans and cottonseeds); mutant varieties of crops; leaves, grasses, and aquatic plants; and highly nutritious fish meal and concentrates.

Stages of Food Processing

Food processing takes many different forms. Some foods require little or no processing, like an apple that can be picked from the tree and eaten immediately. Others are inedible unless they undergo some type of processing. Food processing methods fall into three categories: primary, secondary and tertiary.

- Primary Food Processing: Primary food processing is the process of turning raw agricultural products into foods that can be consumed. In some cases, the food is ready to be consumed once primary processing is finished. An example of this is jerky made from smoked meat. In other cases, primary processing turns the agricultural product into an ingredient that can then be made into a consumable food, such as milling grain to create flour.
- Secondary Food Processing: Secondary food processing is the process of using ingredients produced through primary food processing to create ready-to-eat foods. An example of this is using flour to make dough and then baking the dough to create bread. Other examples include fermenting grape juice with wine yeast to create wine and using ground meat to make sausages.
- Tertiary Food Processing: Tertiary food processing is the large-scale manufacturing of ready-to-eat foods, like frozen pizzas and packaged snacks. The term "processed food" typically refers to food products manufactured through tertiary food processing. Some foods require multiple types of processing to reach their final, consumable forms. For example, tortilla chips. Corn is grown, then harvested and nixtamalized – the process of soaking it in an alkaline solution – before being made into dough. Then, the dough is used to create tortillas, which are then cut and either baked or fried into chips. When this process is done at an industrial scale and sealed in airtight bags, it includes tertiary food processing methods after the food has gone through primary and secondary processing.

Food Processing Methods

There are certain criteria that have to be compiled for the appropriate processing of food, right from the possibility of a pest or bacteria to invade and multiply on foods to the biological activity of foods. The following methods are applied for the proper processing of food:

- Peeling off the outer layers of the raw materials
- Chopping or slicing

- Mincing
- Liquefaction
- Fermentation
- Emulsification
- Cooking
- Mixing
- Gasification such as the addition of a gas in bread or soft drinks
- Proofing
- Spray drying
- Pasteurization
- Packaging

Other Methods of Food Processing

Food preservation is a process involved in food processing employed to prevent the growth of fungi, bacteria, and many other microorganisms. It involves the process of slowing down the oxidation of fats that would lead to rancidity. There are several food preservation methods that are designed specifically to preserve food. Some of the selected few preservation methods are stated below:

- Drying: One of the oldest methods that humans have been and still are using. The only change has come in the way (equipment) food products are dried. Drying is basically reducing the water or moisture content in the product. Hence reducing the favorable conditions for the bacteria to grow and spoil the food product. It is mainly performed to extend the shelf life. Food products can be dried with the application of heat (vacuum dried, tray dried, heat pump dried etc.) or cold temperatures (freeze-dried).

Freeze Dried Foods | Food Processing Methods.

- Fermentation: Fermentation is a food preservation and value addition process for food products. It is the action of microorganisms on food in a controlled and suitable environment. Fermentation process preserves the food with the production of acid or alkali with the introduction of characteristic flavor, hence increasing its shelf life as well as marketability. Various factors such as temperature, type of microbe, oxygen, pH etc. affect this process.
- Sugaring: Sugaring is a processing cum preservation method generally used for fruits such as peaches, plums, pineapple etc. Suitably sized fruit pieces are either cooked with sugar or a dipped in sugar syrup and sterilized with the help of sterilization process.
- Pasteurization: Pasteurization is a processing method that is applied to mainly milk. A mild heat treatment about 72 degrees centigrade for 15 sec is applied and then the milk or the food product is cooled to a temperature of 4 degrees. This time and temperature combination is called High-Temperature Short Time (HTST) pasteurization. Such conditions are enough for the inactivation of microbes with minimal change in nutritive value and sensory characteristics.
- Pickling: Usually, the food is cooked or soaked in chemicals that play a role in the destruction of microorganisms. These include brine (salt and water solution), vinegar, ethanol vegetable oil etc. It's generally practiced for vegetables like cabbage, pepper or raw mangoes. The factors that control the pickling process is temperature, moisture, pH etc.

Pickling Of Vegetables | Food Processing Methods.

- Freezing: Frozen foods are currently in great demand. With this food processing technique now you can enjoy fries, paranthas, chicken nuggets etc in a just a few minutes. Freezing process increases the shelf life of the food products and inhibiting the growth of most of the bacterial species. It is carried out via two processes: mechanical and cryogenic or flash freezing (using liquid nitrogen at -196 to -320 degrees Fahrenheit).

- Food Extrusion: Extrusion is increasingly becoming important in manufacturing products like pasta, ready-to-eat cereals, snacks, pet foods, sausages and much more. An extruder is used as the machine for this processing technique. It consists of a large, rotating screw tightly fitted within a stationary barrel, at the end of which is the die (a tool used for cutting material using a press). A set of mixed ingredients are forced through an opening into the extruder with a design specific and is then cut into a specific shape and size by the blades present. Extrusion process makes mass production convenient, efficient and continuous process, ensuring greater uniformity in the final product.

Food Preservation

The term food preservation refers to any one of a number of techniques used to prevent food from spoiling. It includes methods such as canning, pickling, drying and freeze-drying, irradiation, pasteurization, smoking, and the addition of chemical additives. Food preservation has become an increasingly important component of the food industry as fewer people eat foods produced on their own lands, and as consumers expect to be able to purchase and consume foods that are out of season.

The vast majority of instances of food spoilage can be attributed to one of two major causes: (1) the attack by pathogens (disease-causing microorganisms) such as bacteria and molds, or (2) oxidation that causes the destruction of essential biochemical compounds and/or the destruction of plant and animal cells. The various methods that have been devised for preserving foods are all designed to reduce or eliminate one or the other (or both) of these causative agents.

For example, a simple and common method of preserving food is by heating it to some minimum temperature. This process prevents or retards spoilage because high temperatures kill or inactivate most kinds of pathogens. The addition of compounds known as BHA and BHT to foods also prevents spoilage in another different way. These compounds are known to act as antioxidants, preventing chemical reactions that cause the oxidation of food those results in its spoilage. Almost all techniques of preservation are designed to extend the life of food by acting in one of these two ways.

Low Temperature Preservation

Storage at low temperatures prolongs the shelf life of many foods. In general, low temperatures reduce the growth rates of microorganisms and slow many of the physical and chemical reactions that occur in foods.

Refrigeration

The life of many foods may be increased by storage at temperatures below 4 °C (40 °F). Commonly refrigerated foods include fresh fruits and vegetables, eggs, dairy products, and meats. Some foods, such as tropical fruits (e.g., bananas), are damaged if exposed to low temperatures. Also, refrigeration cannot improve the quality of decayed food; it can only retard deterioration. One problem of modern mechanical refrigeration—that of dehydration of foods due to moisture condensation—has been overcome through humidity control mechanisms within the storage chamber and by appropriate packaging techniques.

Freezing

Freezing and frozen storage provide an excellent means of preserving the nutritional quality of foods. At subfreezing temperatures the nutrient loss is extremely slow for the typical storage period used in commercial trade.

The Freezing Process

The freezing of food involves lowering its temperature below 0 °C, resulting in the gradual conversion of water, present in the food, into ice. Freezing is a crystallization process that begins with a nucleus or a seed derived from either a nonaqueous particle or a cluster of water molecules (formed when the temperature is reduced below 0 °C). This seed must be of a certain size to provide an adequate site for the crystal to begin to grow. If physical conditions are conducive to the presence of numerous seeds for crystallization, then a large number of small ice crystals will form. However, if only a few seeds are initially available, then a few ice crystals will form and each will grow to a large size. The size and the number of ice crystals influence the final quality of many frozen foods; for example, the smooth texture of ice cream indicates the presence of a large number of small ice crystals.

In pure water, the freezing process is initiated by lowering the temperature to slightly below 0 °C, called supercooling. As ice crystals begin to grow, the temperature returns to the freezing point. During the conversion of liquid water to ice, the temperature of the system does not change. The heat removed during this step is called the latent heat of fusion (equivalent to 333 joules per gram of water). Once all the water is converted to ice, any additional removal of heat will result in a decrease in the temperature below 0 °C.

The freezing of foods exhibits a number of important differences from the freezing of pure water. Foods do not freeze at 0 °C. Instead, owing to the presence of different soluble particulates (solutes) in the water present in foods, most foods begin to freeze at a temperature between 0 and −5 °C (32 and 23 °F). In addition, the removal of latent heat in foods during freezing does not occur at a fixed temperature. As the water present in the food freezes into ice, the remaining water becomes more concentrated with solutes. As a result, the freezing point is further depressed. Therefore, foods have a zone of maximum ice crystal formation that typically extends from −1 to −4 °C (30 to 25 °F). Damage to food quality during freezing can be minimized if the temperature of the product is brought below this temperature range as quickly as possible.

Industrial Freezers

The rate at which heat is removed from a food during freezing depends on how fast heat can travel within the food and how efficiently it can be liberated from the surface of the food into the surrounding atmosphere. Industrial freezers remove heat from the surface of a food as rapidly as possible. There are several types of industrial freezers, including air-blast tunnel freezers, belt freezers, fluidized-bed freezers, plate freezers, and cryogenic freezers.

In air-blast tunnel freezers and belt freezers, precooled air at approximately −40 °C is blown over the food products. Packaged foods, such as fruits, vegetables, bakery goods, poultry, meats, and prepared meals, are usually frozen in air-blast tunnels. The packages are placed onto dollies or hand trucks and then rolled into the freezer tunnels. In a belt freezer, food is placed on a conveyor belt that moves through a freezing zone. Bakery goods, chicken parts, and meat patties are frozen using a belt freezer.

Fluidized-bed freezers are used to freeze particulate foods such as peas, cut corn, diced carrots, and strawberries. The foods are placed on a mesh conveyor belt and moved through a freezing zone in which cold air is directed upward through the mesh belt and the food particulates begin to tumble and float. This tumbling exposes all sides of the food to the cold air and minimizes the resistance to heat transfer at the surface of the food.

Plate freezers are used to freeze flat products, such as pastries, fish fillets, and beef patties, as well as irregular-shaped vegetables that are packaged in brick-shaped containers, such as asparagus, cauliflower, spinach, and broccoli. The food is firmly pressed between metal plates that are cooled to subfreezing temperatures by internally circulating refrigerants. Cryogenic freezing is used to freeze food at an extremely fast rate. The food is moved through a spray of liquid nitrogen or directly immersed in liquid nitrogen. The liquid nitrogen boils around the food at a temperature of −196 °C (−321 °F) and extracts a large amount of heat.

Quality of Frozen Foods

Improper freezing or storage of foods may result in detrimental quality changes. When foods with high amounts of water are frozen slowly, they may experience a loss of fluid, called drip, upon thawing. This fluid loss causes dehydration and nutrient loss in frozen food products. During frozen storage, the ice crystals present in foods may enlarge in size, producing undesirable changes in texture. This phenomenon is commonly observed when the storage temperature is allowed to fluctuate. For example, ice cream stored in an automatic defrosting domestic freezer becomes sandy in texture because the ice crystals increase in size as the temperature of the system fluctuates.

Improperly packaged frozen foods lose small amounts of moisture during storage, resulting in surface dehydration (commonly called freezer burn). Frozen meats with freezer burn have the appearance of brown paper and quickly become rancid. Freezer burn can be minimized by the use of tightly wrapped packages and the elimination of fluctuating temperatures during storage.

Thermal Processing and Preservation

The safety and storage life of many perishable foods can be enhanced by the use of high temperatures to inactivate undesirable disease and spoilage-causing microorganisms and to inactivate enzymes in food that can cause spoilage. Three categories of thermal preservation of foods are; Blanching, Pasteurization and Commercial Sterilization.

Blanching

Blanching is a form of thermal processing applied mainly to vegetables and some fruit by exposing them to heated or boiling water or even culinary steam for a short period of time. Blanching is a food processing operation designed to:

- Inactivate enzymes in plant tissues so that enzymatic degradation does not occur in the interval between packaging and thermal processing or during frozen storage or in the early stages of food dehydration and after reconstitution of dehydrated plant foods.
- Wilt vegetable products to enable packing of the products into containers so that proper fill weights can be achieved.

- Drive off inter and intracellular oxygen and other gases from plant tissues so that containers are not deformed by excessively high internal pressures due to expanding gases within the container and to permit formation of a vacuum in the container after thermal processing.

Pasteurization

Pasteurization is a thermal process that involves using temperatures of at least 72 °C for 15 seconds (high temperature short time or HTST process), prior to packaging. Acid food products (pH < 4.6) are mainly pasteurized to inactivate spoilage causing microorganisms. Pathogenic microorganisms cannot grow and do not survive very well in acid foods such as citrus juices or apple juice (with the exception of Escherichia coli 0157:H7).

In low-acid food products such as milk, the basis for preservation by pasteurization is to inactivate pathogenic (disease causing) bacteria and viruses. Many spoilage-causing microorganisms can still survive typical pasteurization process conditions. For example, in milk, the proteolytic and lipolytic bacteria are more heat resistant and can survive the pasteurization process. This explains why the typical spoilage pattern of pasteurized milk reflects the proteolytic (protein degradation) and lipolytic (lipid degradation) action of the psychrotrophic, spoilage-causing bacteria. Because pasteurization does not kill all the psychrotrophic spoilage-causing bacteria in milk, pasteurized milk must be refrigerated to maintain shelf life quality. The durable life date on milk containers reflects the storage life that can be expected when milk is held at 4 °C or lower.

Commercial Sterilization (CS)

The basis for preservation by Commercial Sterilization (CS) is to destroy both spoilage and disease causing microorganisms in low-acid and acid foods, thus rendering the food "commercially sterile". Commercially sterile means the condition obtained in a food that has been processed by the application of heat, alone or in combination with other treatments, to render the food free from viable forms of microorganisms, including spores, capable of growing in the food at temperatures at which the food is designed normally to be held during distribution and storage". Therefore, commercially sterilization involves the destruction of spoilage-causing and disease-causing microorganisms. Commercially sterile foods may contain small numbers of thermophilic bacteria spores; however, the spores cannot germinate and produce actively growing cells at room temperature.

Canning

Canning involves heating the food with a minimum treatment of 121 °C moist heat for 15 minutes. The processes usually involve pre-sealing the food in containers prior to heating. Canning can be traced back to the early 1800's. It is called the "botulinum cook". Today, if a can of food is being sterilized, each food particle must receive the heat treatment (e.g. 121 °C for 15 min). When food is placed in a can, the heat treatment will change since heat transfer to the food takes place at a slower rate. Depending on the size of the can, the time to achieve sterility could be several hours. Most commercially sterile products have a shelf life of 2 years or more.

Ultra-High Temperature Processing (UHT) and Aseptic Packaging

The combination of UHT and aseptic packaging is another form of commercial sterilization. It involves the application of "ultra-high temperature" (heat) to food before packaging, then filling the food into pre-sterilized containers in a sterile atmosphere. This process will render the food shelf stable or commercially sterile without the need for refrigeration. UHT-Aseptic packaging is a relatively new development whereby food can be heated to 140-150 °C very rapidly by direct injection of steam, held at that temperature for short period of time (e.g. 4-6 seconds) and then cooled, in a vacuum chamber to flash off the water added in the form of condensed steam. This is carried out as a continuous flow operation. The decrease in processing time due to the higher temperature, and the minimal come-up time and cool-down time leads to a higher quality product.

The UHT processed food is packaged aseptically into pre-sterilized containers. These are usually cartons made from laminated plastic, aluminium and paper, which are chemically sterilized with a combination of hydrogen peroxide and heat, and then filled in the same piece of equipment which is housed in a sterile environment. There are other forms of packaging that can also be used in aseptic UHT processing: plastic cans, flexible pouches, thermoformed plastic containers, bag-in-box, and bulk totes. UHT-aseptically packaged products have a shelf life of 6 months or more, without refrigeration. It depends on the type of packaging being used. For example, Tetra Pak cartons can eventually be more prone to perforations in the packaging layers, whereas the newer plastic bottles are more resistant to pin whole formation allowing them to have a longer shelf life. Some examples of food products processed with UHT are:

- Liquid products: Milk, juices, cream, yogurt, wine, salad dressings.
- Semi-liquid/solid products: Baby foods; tomato products, fruits and vegetable juices, soups.

Contrary to popular opinion, UHT processed milk and juices do not contain added agents to provide the long storage life at ambient temperature in the laminated cartons. The products are preserved solely through the application of heat. It is critical that the sterilized products are transferred to packaging equipment under aseptic conditions, to avoid contamination after thermal processing. Please note that many products that are UHT treated are not aseptically packaged. This gives them the "advantage" of a longer shelf life at refrigeration temperatures compared to conventional pasteurized products. However, this does not produce a shelf-stable product at ambient temperatures due to the possibility of post-processing recontamination.

Selection of Heat Treatments

The intensity of the heat treatment employed for a particular food preservation application depends upon a number of factors. The main considerations in selecting the required temperature-time conditions for thermal processing are:

- What is the objective or purpose? (Blanching or pasteurization or commercial sterilization)
- Are there additional preservation steps? (Is it combined with other preservation method?)
- What are the physical, chemical properties of the food? (Type of food)
- What is the heat resistance of microorganisms in the food?

Foods that will be consumed within a short period of time after processing can have storage life extended by a combination of pasteurization and refrigerated storage (used for pasteurized milk and for pasteurized, vacuum packaged, cured meats). Longer storage times at ambient temperatures in evacuated sealed containers requires the use of commercial sterilization. The time-temperature combination required for pasteurization and commercial sterilization is determined by the most heat-resistant disease-causing and spoilage-causing microorganisms in the particular food commodity.

For a particular food commodity, the type of thermal processing operation and the rate of heat penetration into the slowest heating portion of the food within a particular container are governed by the food's physical properties (solid vs. liquid, or solid particles suspended in a liquid) and chemical properties (pH, fat content, presence or absence of heat-inducible thickening agents, food components that have protective or antagonistic effects on the thermal resistance of microorganisms). It is imperative that thermal preservation processes be designed so that the slowest heating portion of the food commodity receives the specified time-temperature thermal treatment to minimize risks of illness and/or post-processing spoilage.

The thermal processes applied to foods are governed by the heat resistance of the microorganisms in the food. In low acid foods which are to be thermally processed and vacuum sealed within gas-tight containers, the microorganism of most concern is Clostridium botulinum. The habitat of Clostridium botulinum can be soil (agricultural and forest), water (fresh, brackish and marine) and mud (fresh water and salt water). As a consequence, all foods of agricultural and fisheries origin must be considered as being potentially contaminated with Clostridium botulinum spores.

Thermal preservation treatments of low acid foods that are to be packaged and stored under anaerobic conditions must be designed to ensure the destruction of any Clostridium botulinum spores that may be present in the food with the provision of a wide margin of safety. Actively growing vegetative Clostridium botulinum cells produce a very potent neurotoxin.

Destruction of C. Botulinum Spores

To determine the thermal resistance of heat resistant spores in foods, "Inoculated pack studies" are carried out using a non-pathogenic spore-forming bacterium, Clostridium sporogenes PA3679 (a putrefactive anaerobe). Since PA3679 spores are more heat resistant than those of Clostridium botulinum spores, a process designed to kill PA3679 spores will definitely kill Clostridium botulinum spores with a wide margin of safety.

Determining the Appropriate Temperature and Time Needed

Thermal Death Curves

When microorganisms (bacteria, such as vegetative cells and spores, moulds, viruses and yeasts) are exposed to high temperatures capable of causing death of the organisms, one observes that the population is not killed instantaneously. We see that microbial death during thermal processing follows a logarithmic order. This means that bacteria are killed by heat at a rate that is nearly proportional to the number present in the system being heated.

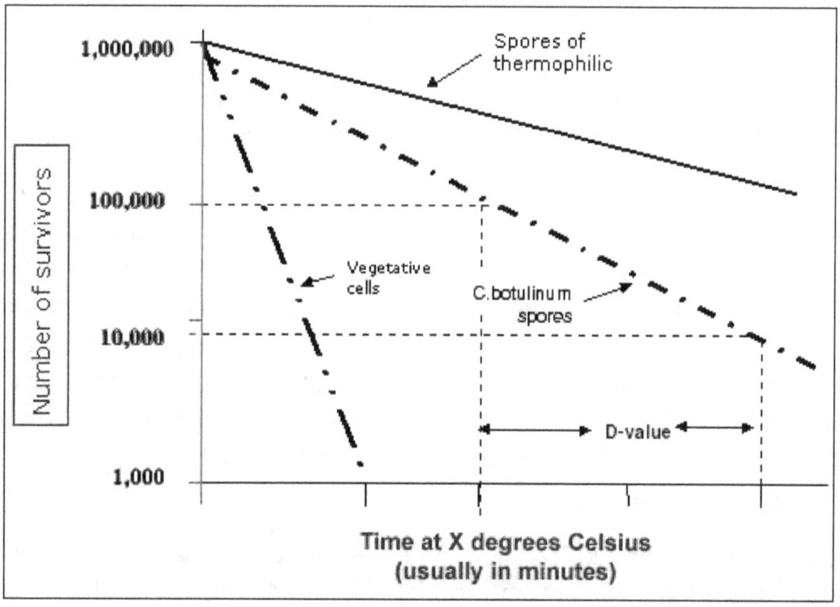

Thermal Death Time Curve.

The survivor curve or thermal death rate curve plotted in the figure depicts the logarithmic order of death. You will note that the time taken to traverse one logarithmic cycle represents the time, at a constant temperature, required to kill 90% of a microbial population. The exposure of microorganisms to a specific temperature for a period of time required to kill 90% of the population is defined as the decimal reduction time or D-value. While the thermal death curve describes the rate of death of a particular microorganism under a specified set of conditions. A thermal death time curve can be constructed from a number of thermal death curves by exposing the microorganism to a variety of temperatures and determining the decimal reduction time at each temperature.

The thermal death time curve provides information about the time required to kill a particular microorganism in a particular food at a variety of temperatures. All of the time-temperature combinations along the plotted thermal death time curve represent the same killing power, with lower temperatures requiring longer time of exposure. You should also note that any point above the line (e.g., 100 minutes at 110 °C) will ensure that the microorganism is killed, while time-temperature combinations that fall below the plotted line (e.g., 100 minutes at 104 °C) represent conditions that will not kill all of the microorganisms present. The thermal death time curve that vegetative bacteria cells have a much lower heat resistance than spores.

Foods that are a combination of solids and liquid components heat up by a combination of convection and conduction heating. Note the location of the cold points in relation to conduction heating (in the geometric centre of a cylindrical container) and convection heating (one third of the way up the centre axis measured from the bottom of the cylindrical container). The cold point of a container of food must receive the required amount of thermal energy to ensure killing of Clostridium botulinum spores that may be present and to ensure a sufficient margin of safety. If a food formulation is changed such that the mechanism of heat transfer is altered from convection heating to conduction heating, and if the processing times are not altered to accommodate the change in mechanism of heat transfer, the food could be under processed and could pose a potential health hazard with respect to botulism.

Canning and Bottling of Foods

Canning is process of preservation by sealing and sterilization of food in air tight container. The main aim of canning is to create aseptic conditions and prevent recontamination. Heat is most common agent used to destroy the micro-organism. Canning process was first invented after prolonged research by Nicolas Appert and further in 1810 Peter Durand of England patented the use of tin-coated iron cans instead of bottles. A typical commercial canning operation may employ the following general processes: washing, sorting/grading, peeling, coring, cutting, container filling, exhausting, container sealing, heat sterilization, cooling, labeling, and storage for shipment. Typical canned products include beans (cut and whole), beets, carrots, corn, peas, spinach, tomatoes, apples, peaches, pineapple, pears and apricots. Juices are also preserved by canning like orange, pineapple, grapefruit, tomato, and cranberry. Canned vegetables generally require more severe processing than fruits because the vegetables have much lower acidity and contain more heat-resistant soil organisms. Methods used for cooking vary widely like in some fruits preliminary treatments steps occur prior to heating but in vegetable preliminary steps done after blanching. Peeling is done either by steam or lye peeling. The principle of canning is same as the process was invented but production and processing of canned foods have changed considerably over the past decades.

Technically bottling refers to storing in glass bottles and canning refers to storing in tin cans, although mode of preservation is same in both.

Preparation For Canning

Selection of Fruits and Vegetables

Fruits and vegetables selected for canning should be fresh. Fruits used for canning should be ripe, firm and evenly matured, and vegetables should be tender. Fruits and vegetables should be stored at cool place and free from dirt, mechanical injury and insect damage.

- Sorting and grading: Sorting for same kind of items is done by passing through series of moving screens. After sorting fruits and vegetables are graded for uniform quality of size and color. Different types of grading machines are used such as screen grader, roller grader, rope and cable grader. Hand picking methods is generally used for soft fruits and berries.
- Washing: Fruits and vegetables are washed through high pressure sprays, steam, by soaking or agitating in water. Vegetables soaked in a dilute solution of potassium permanganate and root crops soaked in chlorine water (25-50 ppm).
- Peeling, Coring and Cutting: Hand peeling, heat peeling, steam peeling, mechanical peeling and lye peeling methods are used. Hand peeling is done in case of irregular shape fruits and vegetables. In heat peeling fruits and vegetables are exposed to a high temperature of 40 °C for 10-60 seconds. In steam peeling, steam is used to loosen the skin which is then removed by mechanical means. Fruit are immersed in 1- 2% lye solution for 30 seconds to 2 minutes in lye peeling. Mechanical peeling is done in case of fruits like peaches, apricots, sweet orange and vegetables like carrot, sweet potatoes. Coring is the process of removing centre of various fruits and vegetables such as apple, pears, pineapple, lettuce and cabbage, done by a hydraulic powered device with turbine wheels. A special blade mounted on the turbine wheel spins and removes the core. Cutting is done according to the requirement of process.
- Blanching: Blanching is done by dipping fruits and vegetables in hot water at temperature 82-90 °C for 2-5 minutes and then immediately cooled by immersion in cold water. Main objective of blanching is to inactivate the enzyme, to soften the texture, to loosen skin, to reduce the number of micro-organism, to enhance the green color of vegetables, to remove acid and astringent taste. Hard water cause toughening of tissue and destroys the natural texture, so it should be avoided.
- Cooling: To keep fruits and vegetable into good condition, cooling is done after blanching.

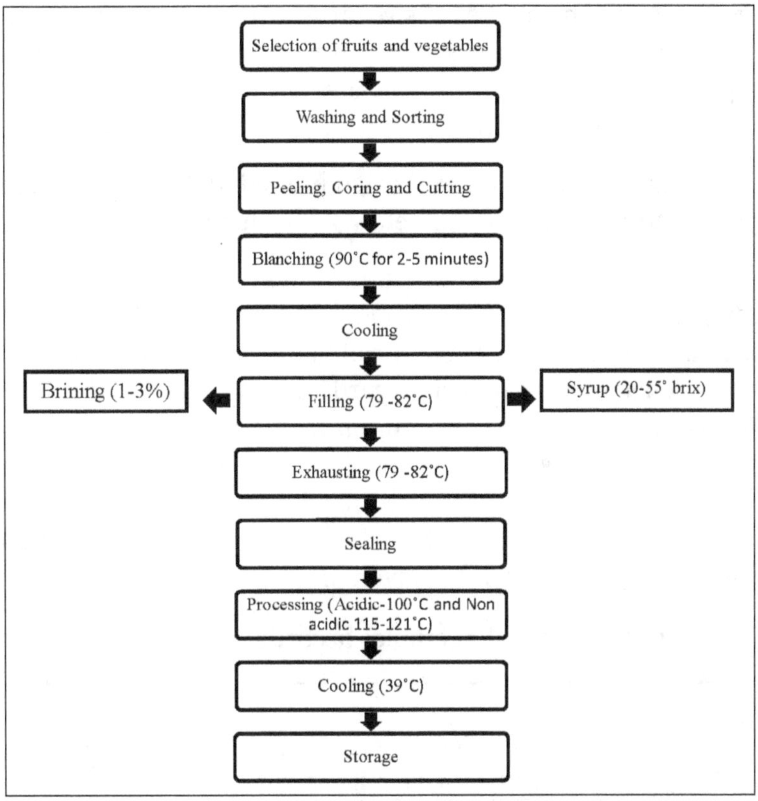

Flow chart of canning.

- Filling: Before filling cans are subjected to steam jet or hot water to remove dust and foreign material. Automatic can filing machine is used by canning industry. Generally for fruits filling is done by hand to prevent the bruising. After filling syrup or brine is added:
    - Syruping: Syrup is a solution of sugar in water, done only for fruits. Syrup is added to improve the flavor and serve as a heat transfer medium for facilitating processing. Strained, hot syrup of concentration 20 to 55° brix is used and filled at about 79 to 82 °C, leaving head space of 0.3 to 0.5cm.
    - Brining: Brine is a solution of salt of concentration 1-3%, used for vegetables. The brine should be filtered through a thick cloth before filling. After Syruping or brining the cans are loosely covered with lids and exhausted. Lidding has certain disadvantages such as spilling of the contents and toppling of the lids. Hence lidding has now been replaced by clinching in which lids is partially seamed.
- Exhausting: Exhausting is the process of removal of air from the cans. It is very essential as it avoids the corrosion of tinplate and pinholing during the storage. Exhausting is done to minimize discoloration, to reduce chemical reaction between the container and the contents. It also helps in better retention of vitamin

and prevents development of excessive pressure during sterilization. For exhausting heating method is generally used but can also be done by mechanical means. The can are passed through a tank of hot water at 82 to 87 °C for 5-10 minutes. At the end of exhausting temperature at the centre of can should be about 79 °C.

- Sealing: After exhausting can are sealed immediately with the help of can sealer and temperature should not fall below 74 °C during sealing.
- Processing: It is heating or cooling of canned foods to inactivate the bacteria. Processing time and temperature should be adequate to eliminate all bacteria and to minimize quality damage. Fruits and vegetables processed at temperature of 100 °C as presence of acid retards the bacteria and their spores. Non-acid vegetables are processed at high temperature 115-121 °C. Temperature and processing time vary with size of man and nature of food. Temperature at the centre of can, should be maintained for long period of time to ensure the destruction of most heat resistant bacteria. Fruits and vegetables are classified into groups depend on pH value given in table.

Table: Classification of Fruits and Vegetables according to pH.

Class	pH	Product
Low acid	Above 5.0	Peas, lima bean, asparagus, cauliflower, potato, spinach, beet, corn, French beans.
Medium acid	4.5-5.0	Turnip, carrot, cabbage, pumpkin.
Acid	3.7-4.5	Tomato, pear, banana, mango, jackfruit, pineapple.
High acid	Below 3.7	Citrus juice, rhubarb, prune, sauerkraut, pickle, chutney.

Processing methods: Generally three types of processing methods are used for canning like open cookers, continuous non-agitating and continuous agitating cookers. Open cookers are made up on wooden tubs or galvanized iron tanks of desired capacity. In continuous non-agitating cookers, cans travel on a continuous moving belt in boiling water crates. Processing time reduced by using continuous agitating cookers.

- Cooling: Cans are cooled rapidly to 39 °C to stop the process and to prevent the stack burning. Cooling is done either by dipping or immersing can in cold water tank or by spraying jet of cold water. In case of canned vegetables, cooling is done by turning in cold water into pressure cooker.
- Storage: After labelling cans are packed in strong wooden cases and stored in cool and dry place. Storage at high temperature should be avoided as it shortens the shelf-life of the product.

Type of Containers

- Tin container: Tin cans are made up of thin steel plate of low carbon content, lightly coated on both sides with tin metal. Coating may not be uniform that result in discoloration of products or corrosion of tin plate. This situation can be avoided by lacquering of cans. There are two types of lacquering:

  - Acid-resistant: It is golden coloured enamel and can coat with it called R or A.R. cans. It is used for high acidic fruits and vegetables e.g., peach, pineapple, grapefruit, strawberry and raspberry.
  - Sulphur resistant: This enamel is also of golden colour and cans are coated with it called C or S.R. cans. It is used for non- acidic products e.g., pea, corn, lima bean, red kidney bean.

- Glass container: Contents are visible in glass container and reusable. They do not contaminate the contents but highly fragile in nature.

Spoilage in Canned Foods

Generally there are two reasons for the spoilage of canned food:

Microbial Spoilage

Thermophilic and mesophilic organism are responsible for canned food spoilage. Thermophilic bacteria can survive at a high temperature of 100 °C. Facultative thermophiles can grow at 43 °C and obligate thermophiles grow at 43-77 °C. Spoilage by mesophilic organism is the indication of under processing. Generally Clostridium, Bacillus, yeast and fungi cause spoilage of can due to formation of carbon dioxide and hydrogen. Thermophiles can cans three types of spoilage:

- Flat sour: Flat sour spoilage occurs mostly in non-acidic foods by Bacillus such as B. coagulans and B. sterothermophilus which produce acid without gas formation. Flat sour spoiled product is unfit for consumption as it has sour odour and highly acidic.
- Themophilic acid spoilage: Clostridium thermosaccharolyticum, an obligate thermophile, is responsible for TA spoilage in which cans swell due to production of carbon dioxide and hydrogen.
- Sulphide spoilage: Also known as sulphur stinker is caused by Clostridium nigrificans in low acid foods and occurs in case of under processing of canned foods.

Spoilage due to Physical and Chemical Changes

- Swell: It is bulging of can due to pressure of gases formed by microbial and chemical action.
  - Hydrogen Swell: Hydrogen gas produced by the action of food acids on metal of can that can cause bulging of can. Canned food remains fit for consumption as it is free of microorganism.

- ◦ Flipper: Swelling of can when both ends become convex and flips out, but when can push back to normal condition by little pressure. Flipper caused by overfilling, under exhausting and gas pressure due to spoilage.
- ◦ Springer: It occurs when the can swells at both end due to insufficient exhausting or overfilling.
- ◦ Soft swell: It is more or less similar to that of flipper in which both ends of can swell. When can is pressed, the end returned back to the normal position and springs back when pressure is removed.
- ◦ Hard swell: It is a final stage of swell in which bulged end cannot be pressed back to normal position and can ultimately burst.

- Leakage: Leakage is due to defective seaming, nail holes caused by faulty nailing of cases, excessive internal pressure, corrosion and mechanical damage during handling.
- Stack browning: When cans are not allowed to cool properly before storing stack browning takes place which results in discoloration, cooked flavor and soft product. So, it is necessary to cool product to 39 °C before storage.

Canning and Bottling of Fruit and Vegetables

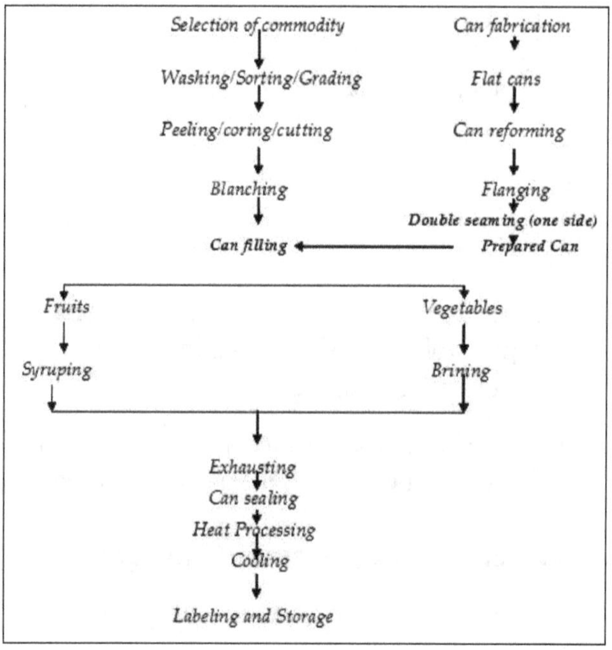

Flow sheet for canning of fruits and vegetables.

Fruit and vegetables are canned in the season when the raw material is available in plenty. Canned products are sold in off season and fetch better returns to the grower as well as processor.

Steps Involved in Canning of Fruits and Vegetables

Preparation of Fruit and Vegetables

Preparation of food commodity for canning consists of washing, sorting, grading, peeling, halving, blanching etc.

Raw Material Selection/Receiving

For canning, fruits should be ripe but firm, evenly matured, free from blemishes, insect damage and malformation. Thus, harvesting at proper maturity is an important step in selection of raw material for canning.

- Most fruits are harvested at soft ripe stage. However, apple, pear, peach and banana harvested at mature stage are preferred for canning.
- Over ripe fruits yield poor quality product, while under ripe/immature fruit generally shrivel or toughen on canning.
- Vegetables except peas, beans etc. are harvested at mature stage to enable them to withstand cooking during sterilization.
- Vegetables like green beans, green peas, ladies finger should be tender and free from soil, dirt etc.
- Tomatoes should however, be firm, fully ripe and uniformly deep red in colour.

The specific requirements for canning of fruits and vegetables are given under Table.

Table: Specific requirement for canning of fruits.

Fruits	Preparation	Syrup Strength (°B)	Exhaust	Processing time at 100 °C A2½ can) minutes	Type of can
Apricot	Use whole or halves, peel by dipping in boiling lye solution (2% NaOH) for 30 seconds to 1 minute, dip in cold water, cut two halves, remove pit, keep immersed in 2% salt solution until filled in can.	40 + 0.1 % citric acid	Exhaust can at 82-100 °C for 6-10 minutes or until temperature in can center reaches 79 °C.	20-25	Plain

Peach	Use whole fruit ,peel by dipping in boiling lye solution (2% (NaOH) for 30 seconds to 1 minute dip in cold water, cut two halves, remove pit, keep immersed in 2% salt solution until filled in can.	40	-do-	25	Plain
Banana	Peel, cut in to slices 12mm thick.	30	-do-	25	Plain
Grape fruit	Dip fruit in hot water at 93 °C-96 °C for 3-5 minutes, remove skin. Dip the peeled whole fruit in 2% hot lye, wash and separate the segments, fill segments in can.	60	Exhaust can at 82-87 °C for 25-30 minutes or until temperature at center of can reaches 79 °C.	30-40	Plain
Guava	Peel, cut into pieces, remove seeds and keep in 2% brine to prevent browning and fill in can.	40	-do-	20	Plain
Litchi	Peel by hand, remove pits.	40 + 0.5% citric acid	-do-	25	Plain
Pineapple	Peel, core and slice fruit, punch circular rings (5.6-8.8 cm diameter)	40	-do-	25	Plain
Pear	Peel by hand, cut longitudinally into two halves, core and keep in 2% brine to prevent browning until filled in cans.	40 + 0.1%citric acid	-do-	20-25	Plain
Jack fruit	Cut fruit into pieces, remove bulbs, separate seed from bulbs, use whole, halves or quarters.	50 + 0.5-0.75% citric acid	-do-	20-25	Plain
Loquat	Cut into halves.	40	-do-	25	Plain
Mango	Peel, cut into slice, dip in 2% salt solution until filled in can.	40 + 0.3%citric acid	-do-	25	Plain
Papaya	Peel, cut into slices, discard seed.	45 + 0.5% citric acid	-do-	25	Plain

Table: Specific requirement for canning of vegetables.

Vegetable	Preparation	Brine strength (%)	Exhaust	Processing (A2½ can) at 121 °C (0.7 kg/ cm² steam pressure)	Type of can
Asparagus	Wash, grade, cut lengthwise into pieces of desired size, blanch for 2-3 min.	2.25%	Exhaust can at 90-100 °C for 7-10 minutes, (79 °C in can center)	24	Plain /sulphur resistant
Beans	Use tender, stringless beans, slice 2.5 cm length, blanch and fill in can.	2.25	-do-	40	Plain/ sulphur resistant
Cabbage	Shred cabbage, blanch in 1% citric acid boiling solution for 5-6 min, cool in 2% brine prevents discolouration fill in cans	2.0	-do-	40	Plain
Cauliflower	-do-	2	-do-	20	Plain
Carrot	Wash, scrape and blanch for 5-10 minutes in boiling water	2	-do-	25	Plain
Mushroom	Use tight buttons, blanch for 4-5 minutes, immerse in 2% salt solution, fill in cans.		-do-	25	Plain
Peas	Shell, grade, boil for 3-5 min.	2 + 2.5% sugar	-do-	45	Sulphur resistant
Potato	Peel, keep in 2% brine, blanch in boiling water for 2-3 min, keep in 2.5% $CaCl_2$ for 1 hour, wash and fill in cans.	2	-do-	45	Plain
Okra	Tender whole, blanch in boiling water for 1-2 minutes, cool in brine solution of 1-2% for 10 minutes, fill in can.	2 or in tomato sauce	-do-	35	Plain

Washing

Fruit and vegetables are generally washed with water to remove dust, dirt and adhering surface microflora. Fruits like peach, apricot etc. are lye peeled so not washed before peeling. On the other hand, washing after peeling removes vitamins and minerals and should be discouraged. Different methods of washing include soaking or agitating in water, washing with cold or hot water sprays etc.

- Mechanical washers involve agitating or tumbling the commodity on moving belts or revolving screens while they are immersed in water or subjected to water sprays.
- Washing by using high pressure sprays is most satisfactory.
- Detergents are frequently used in the wash or rinse water.
- Vegetables may be soaked in dilute solution of potassium permanganate or chlorine (25-50 ppm) for disinfection.
- The water temperature should be kept low to keep the fruit firm and to reduce leaching losses.
- High pressure sprays should not injure the fruits.

Bacteria and other contaminants can accumulate in the wash water and hence appropriate cleaning and chlorination practices be followed.

Sorting and Grading

Sorting and grading ensures the removal of inferior or damaged commodity. For sorting, inspection belt can be used, in addition to trained personnel who detect poor quality produce unsuitable for canning.

- Automatic colour sorters can be used for sorting to reduce labour cost.
- The fruit and vegetables are graded to obtain uniform quality with respect to size, colour etc. after preliminary sorting.
- Grading can be done either manually or with the help of grading machines.
- For mechanical grading, the fruit and vegetables are passed over screens with holes of different diameter.
- Different types of mechanical graders include screen grader, roller grader, rope or cable grader etc. Screen graders fitted with vibrating screens of copper with circular openings are most commonly used. A set of six screens is generally provided to accommodate different sizes.
- Soft and berry fruits are generally graded manually.

- Plums, cherries and olives are graded whole while peaches, apricot, pears, mangoes etc. are graded after cutting them into halves or slices for canning.
- White button mushrooms are graded on cap size basis. Only healthy and light buttons with cap diameter up to 2.5cm and compact head are graded as A grade while, cap diameter beyond 2.5cm as B grade.

Peeling, Coring and Pitting

These are the primary unit operations for preparing fruit and vegetables for canning. Depending upon the type of commodity, peeling and coring methods are selected such as; (1) by hand or knife (2) by machine (3) by heat treatment (4) by using lye solution. Cores and pits in fruits like apple, peach, apricot etc. are removed by hand or by machine (de-corer).

Table: Common methods for the peeling of fruit and vegetables.

S.No.	Peeling method	Commodity
	Knife peeling (manual)	All fruit and vegetables
	Mechanical knife peeling	Apple, pear, pineapple
	Hot water peeling	Tomato
	High pressure steam/water peeling	Potato, tomato
	Abrasive peeling	Potato, ginger, carrots
	Flame peeling	Brinjal, onion, garlic
	Lye peeling	Orange segments, peach, apricot, nectarines, pears

- Peeling by hand: Many fruit and vegetables are peeled and cut by hand with the help of peeling knives. The peeling knife with a curved blade and a special guard to regulate the depth of peeling can be used for uniform peeling in case of irregular fruit shapes.
- Mechanical peeling: Mechanical peeling, coring and cubing machines are used for peeling pears, apples, carrots, turnip, potatoes etc. Similarly, automatic peelers are used for peeling of peaches and cherries.
- Mechanical/Knife peeling: Mechanical knife peelers are used for peeling of fruits like apples and pears. In mechanical knife peeler either stationary blades are pressed against surface of rotating food commodity or the rotating blades are pressed against the stationary food to remove the skin.
- Abrasive peeling: It is used for peeling potatoes, ginger, carrots etc. The food commodity is fed on to the carborundum rollers or placed into a rotating bowl which is lined with carborundum crystal acting as abrasive surface. With the continuous supply of water, the rotating abrasive surface removes the skin from the surface of the food.

- Flame peeling: Flame peeling is used in onions, garlic and brinjal. The peeler consists of a conveyor belt which carries and rotates food through a furnace heated to more than 1000 °C. The outer layer and root hairs of onion are burnt off and charred skin is removed manually.
- Peeling by heat or hot water: In this method peaches and potatoes are scalded in steam or boiling water to soften and loosen skin, which is subsequently removed manually. Infra-red heat peeling can also be used for peeling of apples and tomatoes.
- Flash steam peeling: In flash-steam peeling, the fruit and vegetables are fed into a slow rotating (4-6 rpm) pressure vessel. High pressure steam (1500 kPa) is then introduced into the rotating vessel to expose all food surfaces to the steam for specified period depending upon the type of fruit. When the pressure is instantly released, the steam formed under the skin causes the surface of the food to flashes off. Most of the peeled material is discharged with the steam and finishing is done with additional water sprays to remove any skin traces.
- Lye peeling: Lye is a boiling aqueous solution of caustic soda (Sodium hydroxide) or Potassium hydroxide (1- 2%) used in conjunction with ample water supply and heat source for peeling. Fruit and vegetables like peaches, nectarines, apricot, sweet orange segments, carrots and sweet potatoes are peeled by dipping them in boiling caustic soda (1-2%) for 1-2 minutes (depending upon the strength of lye, temperature/maturity and nature of fruit or vegetable) followed by dipping in cold water. The hot lye loosens the skin from the flesh underneath which is removed by gentle rubbing of fruit by hand. The fruit can also be dipped in a dilute solution of hydrochloric acid or citric acid for few seconds to neutralize the alkali. The method is very quick and efficient to reduce wastage and peeling cost. The effectiveness of lye peeling depends upon lye concentration and temperature, product holding time and agitation.
    - Lye peeling equipment varies from simple stainless steel (SS) pan for lye solution with SS baskets as cages for holding the food commodity to fully automatic system.
    - In cottage and small scale canning units, the peeling system consists of three SS tanks attached in series, the one of which is having provision for steam, the second tank contain dilute solution of citric acid or hydrochloric acid while the third is filled with tap water.
    - The fruit or vegetables placed in perforated SS crates/basket or cages are dipped in the first tank which contains boiling hot lye solution. After 1-2 minute of dipping treatment, the crates are immediately dipped in second tank to neutralize the lye and final washing is carried out in third tank.

Cutting/Halving/Slicing

After peeling, the fruits are halved or cored either manually or mechanically. However, peeled fruit should always be kept submerged in either water, containing 1-2 % salt solution or acid to avoid enzymatic browning. Peaches, apricot, pears, tomatoes etc. are peeled before canning. However, the fruits which are canned retain better nutrients as compared to peeled fruits.

Blanching

Treatment of fruit and vegetables with boiling water or steam for short periods followed by immediate cooling prior to canning is called blanching. The basic objectives of blanching are as under:

- To inactivate enzymes.
- To clean the product initially to decrease the microbial load and to preheat the product before processing.
- To soften the tissue to facilitate compact packing in the can.
- To expel intracellular gases in the raw fruit to prevent excessive pressure built up in the container.
- To allow improved heat transfer during heat processing.
- To ensure development of vacuum in the can and to reduce internal can corrosion.

Blanching is carried out either by hot water or using live steam. Water blanching is generally of the immersion type or spray type as the product moves on a conveyer. Only soft water should be used for blanching as hard water toughens the tissue and destroys the natural texture.

Prevention of Browning

Some fruits which cannot be blanched due to their delicate tissue structure are treated with some chemicals to prevent oxidative browning, occurring due to exposure to oxygen during peeling and slicing. Oxidative browning is caused by action of oxidase enzyme with catechol and tannins and is common in peach, apple, potato, mushroom, cherry, apricot, grapes and persimmon. Pineapple, tomato and melons are however not prone to browning. Common methods used to prevent browning are as under:

- Sulphite treatment: Fruits are dipped in a solution containing 2000-4000 ppm SO_2 for 2-5 minutes. SO_2 fumigation can also be used commonly for grapes dehydration.
- Acids: Common acids used to increase acidity include citric, fumaric, tartaric,

acetic, phosphoric etc. Low pH of solution is known to act as inhibitor for enzyme polyphenol oxidase thus inhibits the browning of fruits. The peeled fruits, slices or cut surfaces are dipped in a 1-2 % citric acid solution to prevent browning.

- Antioxidants: Ascorbic acid is commonly used as an antioxidant in most canned fruits. It acts as an inhibitor of peroxidase in some fruits like kiwi fruit. It also reduces quinones, which are generated by polyphenol oxidase upon oxidation of polyphenols to phenolic compounds thus preventing their conversion to brown pigments. Ascorbic acid can be used as such or mixed in dry sugar, citric acid or in syrups.
- Sugars: Sugar syrup is used to prevent browning in peeled and sliced fruits by inhibiting oxidation by partially excluding air in the tissues. Sugar is mixed with ascorbic acid and citric acid as an effective agent against loss of texture, colour and flavour. Addition of chitosan in filtered apple and pear juices also prevents enzymatic browning.
- Salt: Dipping of peeled and sliced fruit and vegetables in 1-2% salt solution also prevent enzymatic browning, as salt acts as inhibitor for polyphenol oxidase.

Filling in Cans

Tin cans are washed in hot water or in steam jet to remove any adhering dust or foreign matter. The cans are then sterilized by dipping in hot water tank or the cans are passed through a steam sterilizing tunnel before use. Generally plain cans are used however, for coloured fruits like plums, black grapes; strawberries etc. lacquered cans are employed. The fruit and vegetable either slices, halves or whole are filled into the cans keeping in view the declared drain weight.

Syruping or Brining

The cans are filled with hot sugar syrup (35-55%) for fruits and with hot brine (2-10%) concentration for vegetables. The purpose of syruping or brining is to help in transfer of heat within the food pieces during processing. It also improves the taste of the canned product, fill up the inter-space between the fruit or vegetables in the can. The syrup or brine is added to the can at a temperature of 79-82 °C, leaving 0.32-0.47cm head space either manually or in automatic machines. In automatic machines, the prepared syrup or brine is drawn into the cans through a horizontal pipe having a row of small holes. The cans travel on a continuous belt in an inclined position below the syrup or brine pipe and get filled, the overflowing excess syrup is pumped back into the syrup tank by a centrifugal pump.

Exhausting

Exhausting is a unit operation in which practically all air from the contents in the can is removed before sealing. The purpose of exhausting and creation of vacuum is to create an anaerobic environment in the can that would inhibit microbial spoilage. The removal of air from the contents also reduces the risk of corrosion and pin holing of the tin plate and discoloration of can contents. Exhausting helps in better retention of vitamin C. Expansion and shriveling of contents during heating help to avoid over filling or under filling of the can. (Corn and peas expand on boiling in brine while strawberries shrivel upon heating in sugar syrup). The vacuum in can prevents bulging of the can during storage at higher attitudes or in hot climate. It also prevents excessive pressure and strain during sterilization.

Methods of Exhausting

There are generally three methods of exhausting the cans to remove headspace gas and creation of vacuum:

- Heat/thermal exhausting: Heat exhausting is used in cans. The can covered with the lid or loosely sealed or clinched is passed through a tank of hot water at about 82-87 °C or on a moving belt through a covered steam box. In water exhaust box, the cans are placed in such a manner that the level of water is 1.3-2.5 cm below their tops. The time of exhausting varies from 5 to 25 minutes depending upon the nature of the product. At the end of the exhausting, the temperature at the centre of the can should be about 79 °C. During exhausting, the steam replaces the air inside the can and it is sealed while still hot.
- Steam flow or steam-vacuum closing: In this system, high steam pressure is injected into the can headspace (at 100 °C for 5-8 minutes) just prior to closing. Thus, all the air inside the can is quickly replaced with steam, which will condense and form vacuum following seaming. Steam vacuum closure coupled with hot fill, assures very high vacuum in the can.
- Mechanical vacuum sealing: In high speed mechanical vacuum sealing, the cans filled with the product and covering syrup or brine, are passed through a clincher that clinches the cans (first operation roll seam) but does not form an airtight seal. The cans are subjected to a vacuum for a short period of time to remove the free headspace air but not all dissolved gases within the product. However, during this process some syrup may be drawn along with the dissolved air. To avoid syrup spillage, a pre-vacuuming step before vacuum closing is recommended. High vacuum closing is also used in case of glass jars where the jar is placed in a closed chamber in which high vacuum is maintained.

Seaming/Closing

Immediately after exhausting, the cans are sealed by using a double seamer. Double seaming is a two-step operation. In the first operation, the can lid is inserted on the can body hook by holding and rotating the lid-in-position can between two rollers. This operation is called as clinching; during which first operation roller gently guides the lid in the body hook. The next step is to press the seam using the second operation roller, which results in an appropriate overlap of the body hook and cover hook which results in an appropriate countersink. Between the cover hook and body hook lays a layer of sealing compound which ensures the sealing process. The critical parameters for an ideal hermetic seam are body hook, cover hook, seam thickness, seam width and overlap which need to be carefully controlled to prevent leakage in the can.

Immediate closing of the cans is required to prevent excessive cooking of the surface of the product. Double seamers are of different designs and capacities like hand operated, semi-automatic and fully automatic. Modern double seamers operate at high speeds (300 cans per minute) while liquid products are sealed in cans at speed of up to 1600 per minute.

Coding/Embossing

Coding of lid of the can is essential to identify the can, once it is closed. The code provides the necessary information about the product like name of canning unit, product packed in the can, date of packing; lot number etc. Coding is done on the second lid (end cover) of the can just before sealing.

Heat Processing

The cans after sealing are immediately transferred to the heating retorts to achieve sterilization of contents. Heat processing consists of heating cans to a predetermined time and temperature of heating to eliminate all possibilities of microbial spoilage. Over cooking should be avoided as it spoils the texture, flavour and appearance of the product. Generally all fruits and acid vegetables can be processed satisfactorily in boiling water (100 °C) as the presence of acid retards the growth of bacteria and their spores. While non-acidic vegetables (except tomato and rhubarb) are processed at higher temperatures of about 115-121 °C under pressure. It needs to be ensured that required temperature reach the centre of the can. The temperature at the centre of the can should be maintained for sufficiently long period to destroy spores of more heat resistant bacteria.

Processing Methods

Processing methods differ with the kind of fruit and vegetables to be processed. The cans containing most fruit and acid vegetables ($pH < 4.5$) are heated in open cookers, continuous non-agitating cookers and continuous agitating cookers.

- Open cooker consists of stainless steel (SS) or galvanized iron tanks to which perforated water pipes are placed underneath the false bottom to supply the steam for heating of water. The sealed cans are placed in SS or GI crates and immersed in the tank containing boiling water.
- In continuous non-agitating cookers, the cans travel in boiling water in crates carried by over-head conveyors on a continuous moving belt.

- While in continuous agitating cookers, the sealed cans while moving on the belt are rotated by a special mechanical device to agitate the contents of the cans. This helps in reducing the processing time.
- For low acid foods like vegetables (pH > 4.5), with hard texture, the processing is carried out in a pressurized vessel (retorts) at elevated temperatures (= 110 °C) under higher steam pressure (2-3 atmospheres). The retorts vary in shape and size (horizontal or vertical), type of operation (batch to continuous, non-agitating to agitating) and with different types of heating media such as water, steam, steam/air or flame.
- In small scale canning units, vertical stationary retorts are generally used. They are made of cast iron cylinders and are fitted with a lid which can be bottled steam tight. They are provided with steam and water feeds, drain cock, safety valve, pressure gauge and thermometers.

Factors Effecting Sterilization

Altitude

As boiling point of the water decreases with the increase in altitude, the processing time for a particular product standardized at sea level should be increased with the increase in altitude.

- For every increase of 152 meters above mean sea level (a.m.s.l.) in altitude, the boiling point of water decrease by about 1 °C and normal processing time has to be increased by about two minutes.
- Similarly, as the altitude above mean sea level increases, the pressure required to maintain the specified processing temperature also increases.
- It is necessary to maintain higher gauge pressure if processing is carried out at altitude greater than 305 meter a.m.s.l. (1000 ft a.m.s.l.).

pH

Inherent pH value of the food commodity greatly influence the processing schedule to achieve required destruction of micro-organisms. The lower the pH, the greater is the ease with which a product can be processed or sterilized. On the basis of pH, the foods are classified into following four distinct classes:

- Low acid foods (pH = 5.0): The low acid foods are the class of foods which require greater care in its preservation as they are subjected to spoilage by thermophilic and mesophilic putrefactive anaerobes including Clostridium botulinum. These foods include most of the vegetables like peas, beans, corn, asparagus, mushroom and corn etc. They are necessarily required to be processed above 110 °C temperatures and under steam pressure.

- Medium acid foods (pH 5.0-4.5): Medium acid foods like meat and vegetable mixtures, spaghetti, soups and sauces also need processing above 110 °C under steam pressure. They are subjected to spoilage by thermophilic and mesophilic anaerobes. They also include thermophilic anaerobes not producing hydrogen sulphide but causing flat sours.
- Acidic foods (pH 4.5-3.7): Most of the fruits including peas, fig, pineapple, nectarines, mango, apple, tomato, subject to spoilage by non-spore forming aciduric, butyric anaerobe like Clostridium pasteurinum and thermophilic anaerobe are classified as acidic foods. They can be processed in boiling water at temperature of 100 °C.
- Highly acidic foods (pH = 3.7): Highly acidic foods like Sauerkraut, berries, citrus juices, grapefruit, rhubarb and pickles are included in this group. As the bacterial spores do not germinate and grow at pH values below 4.5 and do not harm the product even if they are not destroyed in canned fruits. A canned product having pH value of less than 4.5, can be processed in boiling water at 100 °C, but a product with a pH value above 4.5 requires processing at 115-121 °C under pressure of 0.7 to 1.05 kg cm^{-2} (10-15 pounds inch-2).

Cooling

Immediately after processing, the sealed cans are rapidly cooled to approximately 35-40 °C to stop the cooking process and to prevent stack burning. Prolonged heating results in an inferior and uneven product, like peaches and pears become dark in colour, tomatoes turn brownish and become bitter in taste while peas becomes mushy with a cooked taste. Cooling is done by immersing or passing hot cans in cold water tanks, by spraying cans with jet of cold water, by passing cold water in to a pressure cooker or exposing cans to air. Water used for cooling should be non-corrosive, low in bacterial and yeast count and chlorinated with 2 ppm of available chlorine.

Storage

After cooling, the cans are stacked to allow the outer surface to dry, as even a small traces of moisture are likely to induce rusting. The cans are then labelled either manually or by using labeling machine and packed in wooden or corrugated cartons. They should be stored in cool and dry place (below 30 °C). Adequately processed cans usually ensure an acceptable product quality up to at least one year. Storage of cans at higher temperatures should be avoided to prevent the risk of thermophilic growth and spoilage.

Bottling of Fruits and Vegetables

Bottles have proved to be very good containers for home preservation of fruits. Although their initial cost is high, they can be used several times and last for many years if carefully handled. The fruits look attractive through the glass and do not develop metallic flavour. Bottling does not need a sealing machine but is not suitable from the manufacturer's point of view as the initial capital required is high. There are many types of glass containers of different shapes and sizes and with various types of hermetic seals. Jars fitted with wire clamps are considered to be the best. The products remain in a very hygienic condition and do not come into contact with rubber or metal.

Glass Containers

Glass containers are chemically inert, clear, transparent, rigid, resist internal pressure, heat resistant and are cheap in comparison to other packaging materials. Glass containers are the excellent barriers to solid, liquids and gases. They preserve food against odor and flavour contaminations. But when faulty closures are used odour and flavour contamination may occur, hence; the closure should be air tight. Glass does not deteriorate with age in comparison to other packaging material but are light in weight and are fragile (breakable) with thermal shock and impact.

Types Of Glass Containers

- Bottles: The bottles have narrow neck and small closure over the top. Narrow neck facilitates pouring and reduces the size of closure. Bottles are used for packing liquids and small sized solids.
- Jars: They do not have any appreciable neck but are wide mouthed bottles. They are easy to clean and easy to take out product from them. They are used for packing jams and powders.
- Tumblers: They are similar to jars but do not have any neck and no finish. They can be used for packing jams and jellies.
- Jugs: These are large sized bottles with carrying handlers. Used for packing liquids, foods in large quantities like ½ gallon or more.
- Vials: These are small glass containers. Vials used for packing pharmaceuticals while ampoules are used for packing small quantity like spices, food colors, aroma, essences etc.

Properties of Glass Containers

- Glass containers are chemically inert: Almost all types of chemicals can be packed in the glass containers except hydrofluoric acid (it eats the glass hence packed in plastic container). Oils and fats have no reaction with the glass. Water has little or negligible reaction with the glass at low temperature. The products like drugs and transfusion liquids are packed in specially treated glass containers as at higher temperature (during sterilization) the hydrogen from the water is displaced.

- Clarity of glass containers: Products packed in glass containers are easily visualized from outside especially useful for the products kept on the shelves. But some nutrients are packed in colored bottles e.g. brown bottles, amber colored bottles. Opal glass is a ground glass, in which visibility is lost. Gin or opal glass is also used in bathrooms.
- Glass containers are rigid containers:
    - Rigidity helps in filling of containers.
    - Make it possible for stacking.
    - It also helps in loading and vacuum filling of containers. It also provides support to the material.
- Glass containers resist the internal pressure: Glass containers offers resistant against internal pressure brought about by CO_2 or other gases in the product e.g. beer, beverages, soft drinks etc.
- Heat resistant: Glass containers are heat resistant in comparison to other packaging material except cans. A temperature of 1500 °C is applied during manufacturing of glass. Melting point of alumina is 2000 °C, which is used for making glass. Viscous hot materials are poured into glass containers while they are still hot.

Disadvantages of Glass

- Glass containers are fragile/brittle, hence great care is required to be exercised during handling.
- Glass containers are heavy in weight.
- Glass containers are not easy to dispose.

Closures

Closures should prevent loss of contents and must make reseal (crown corks do not make reseal). Closure should not react with the product e.g. in ketchup, formation of black neck takes place. Different types of closures are Crown corks, roll on cap, lug cap, vacuum seal, temper proof, snap fit, press on caps and screw caps.

Procedure for Bottling

- The bottles are thoroughly washed and sterilized.
- The fruit slices are filled leaving about 3 cm space at the top of the jar or bottle.
- The sugar syrup recommended for different fruits is filled boiling hot leaving a head space of 1-1.5 cm.

- Exhausting and sterilization: Separate exhausting of bottles is not required and it is done simultaneously with sterilization by putting a pad of cloth (false bottom) under the bottles.
- The bottles should not be abruptly immersed in hot water, otherwise they may break because of sudden rise in temperature. The temperature of the water should be about the same as that of the contents and should be raised gradually and the bottles kept in the boiling water for the required time.
- At the start of sterilization the lids are left loose and the level of boiling water should come up to the neck of the bottle.
- When sterilization is over, the mouths of bottles and jars should be immediately closed or corked tightly.
- Cooling of bottles is done and the bottles are labeled after drying. The products preserved in bottles require more attractive labels.
- Store in cool and dry place.

Thus, canning and bottling is a well-tested acceptable method of preservation of fruit and vegetable for conversion into stable products. For canning no chemical preservative is used. Shelf-life of adequately processed product in cans is around one year. Keeping in view the versatility of the process, canning and bottling of fruit and vegetable can be adopted as a successful enterprise.

Fermentation and Pickling

Although microorganisms are usually thought of as causing spoilage, they are capable under certain conditions of producing desirable effects, including oxidative and alcoholic fermentation. The microorganisms that grow in a food product, and the changes they produce, are determined by acidity, available carbohydrates, oxygen, and temperature. An important food preservation method combines salting to control microorganisms selectively and fermentation to stabilize the treated tissues.

Pickled Fruits and Vegetables

Fresh fruits and vegetables soften after 24 hours in a watery solution and begin a slow, mixed fermentation-putrefaction. The addition of salt suppresses undesirable microbial activity, creating a favourable environment for the desired fermentation. Most green vegetables and fruit may be preserved by pickling. When the pickling process is applied to a cucumber, its fermentable carbohydrate reserve is turned into acid, its colour changes from bright green to olive or yellow-green, and its tissue becomes translucent. The salt concentration is maintained at 8 to 10 percent during the first week and is increased 1 percent a week thereafter until the solution reaches 16 percent. Under properly controlled conditions the salted, fermented cucumber, called salt stock, may be held for several years.

Salt stock is not a consumer commodity. It must be freshened and prepared into consumer items. In cucumbers this is accomplished by leaching the salt from the cured cucumber with warm water (43–54 °C [110–130 °F]) for 10 to 14 hours. This process is repeated at least twice, and, in the final wash, alum may be added to firm the tissue and turmeric to improve the colour.

Pickled Meat

Meat may be preserved by dry curing or with a pickling solution. The ingredients used in curing and pickling are sodium nitrate, sodium nitrite, sodium chloride, sugar, and citric acid or vinegar. Various methods are used: the meat may be mixed with dry ingredients; it may be soaked in pickling solution; pickling solution may be pumped or injected into the flesh; or a combination of these methods may be used. Curing may be combined with smoking. Smoke acts as a dehydrating agent and coats the meat surfaces with various chemicals, including small amounts of formaldehyde.

Deterioration of Fermented and Pickled Products

Fermented foods and pickled products require protection against molds, which metabolize the acid, developed and allow the advance of other microorganisms. Fermented and pickled food products placed in cool storage can be expected to remain stable for several months. Longer storage periods demand more complete protection, such as canning. Nutrient retention in fermented and pickled products is about equal to retention for products preserved by other methods. Carbohydrates usually undergo conversion to acid or to alcohol, but these are also of nutritive value. In some instances, nutrient levels are increased because of the presence of yeasts.

Food Irradiation

The term “food irradiation” may be applied to any process that exposes food either to electromagnetic radiation or to high-energy particles. Electromagnetic energy can be generated by radioactive isotopes, as in the case of gamma ray irradiation, or by bombardment of thin metal films with high-energy electron beams to produce radiation, as in the case of X-ray irradiation. Alternatively, a high-energy electron beam (“e-beam”) can be directed at the food itself. In all of these cases, the radiation is absorbed by the food, and more particularly, by the microbial organisms in the food. This absorption disrupts the complex organic molecules of the microbes, either preventing the microbes from reproducing or killing them outright. The effectiveness of the treatment varies based on the type of radiation used, the intensity of the radiation, and the microbe in question.

Gamma Ray Irradiation

The simplest form of irradiation, at least in concept, is gamma ray irradiation. In this form of irradiation, the source of radiation is a radioactive element that emits photons in the gamma ray range of the electromagnetic spectrum. Gamma ray photons have a higher frequency (and therefore energy) than either ultraviolet or X-ray photons. Gamma rays can penetrate a target food (or medical product) to a depth of several feet and reach microbial contaminants anywhere within that range.

While simple in concept, gamma ray irradiation can be difficult in practice. The first difficulty is selecting a radioactive source element. In addition to radiating gamma rays, many radioactive elements also produce alpha rays (helium nuclei), beta rays (high-energy electrons or positrons) and/or high-energy neutrons. Alternatively, they might decay into another radioactive substance that generates these other forms of radiation. The other forms of radiation are undesirable because they have the potential to make the target food (or medical product) radioactive. To date, the only radioactive isotopes approved as having the proper radiation profile are Cobalt 60 and Cesium 137, with only Cobalt 60 being actually used for food irradiation at the present time. These radioactive isotopes are produced by exposure of the ordinary element to a nuclear reactor core, and their availability may be conditioned on the continued availability of nuclear power.

Even after a source is selected, there are logistical complications in gamma ray irradiation. Radioactive elements do not have an "off" switch, nor do they come equipped with directional or intensity controls. Gamma rays can be contained by immersion of the source in a sufficient quantity of water, but the source must be removed from the pool in order to irradiate the target food. In order to prevent inadvertent gamma ray exposure, the source must be insulated from the outside world by several feet of concrete.

E-Beam Irradiation

E-beam irradiation, though it uses the same term as gamma ray irradiation, is a completely different kind of treatment. High-energy electron beams are produced in an electron gun, a larger version of the cathode ray gun found in devices such as televisions and monitors. The electrons can be directed by a magnetic field to a target food. The term "irradiation" is really a misnomer, since the food not exposed to electromagnetic radiation or beta rays (electrons produced by a radioactive source). Nevertheless, the process has a similar effect to that of gamma ray irradiation. E-beam irradiation requires shielding as well, but nothing like the concrete bunkers used in gamma ray irradiation. The disadvantage of the e-beam is its short penetration depth (about an inch), preventing its application to many foods and limiting the amount of food that can be processed in bulk.

X-Ray Irradiation

X-ray irradiation is a relatively new technique that combines many of the advantages of the other two methods. Like gamma ray irradiation, X-ray irradiation consists of exposing food to high-energy photons with a long penetration depth. In this case, however, bombarding a metal film with a high energy electron beam produces the photons, allowing the radiation to be turned on and off. The device is a more powerful version of the X-ray machines used in medical offices. The device still requires heavy shielding, although the amount of shielding required is less than that for gamma ray irradiation. No radioactive substances or by-products are used in, or result from, the process.

Regardless of form, food irradiation is fundamentally about how much energy is absorbed by the target food. It is helpful to have a measurement for what dosage of radiation will be required independent of the amount of food to be irradiated. For this reason, radiation doses are measured in kiloGray (kGy). A dosage of one kGy indicates that the target sample receives 1000 Joules (metric units of energy, abbreviated J) per kilogram of sample mass. When measuring the effect of radiation on the microbe population of food, it is useful to have a measurement that does not depend on the number of microbial organisms in a particular sample of food. For this reason, the effect of radiation on microbes is measured by a dosage called the D value. The D value is the dosage of radiation required to reduce the microbe population of a sample by 90%.If a particular organism has a D value of 0.5 kGy in a particular kind of food, then exposing a 1 kg sample of that food containing the organism to 500 J of radiation will kill 90% of the population of that organism. An additional amount of dosage equal to the D factor will reduce the remaining microbe population by 90%. Thus, exposing the sample in the example above to 1000 J of radiation would reduce the microbe population by 99%; 1500 J would remove 99.9% of the microbe population; etc. Varying the power of the source or the duration of exposure controls the amount of radiation the target receives. The energy of electron guns used for e-beams and X-rays is typically measured in electron volts (eV), units of energy convertible to J.

The D value will depend primarily on the type of food irradiated and the type of organism to be eradicated by irradiation. Generally, the more complex the organism, the more sensitive the organism will be to radiation, since the operation of complex microbes is easier to disrupt. Viruses, the simplest form of life, are most difficult to destroy. Many bacteria collapse into a dormant state known as a spore (as contrasted with the vegetative state) when conditions are unfavorable to growth (e.g., when the oxygen or temperature levels are too low). The D value for spores is higher than the corresponding D value for the vegetative state.

Other factors that affect the D value are the strain of organism involved, the state of the food (frozen or unfrozen), ambient oxygen and temperature. The difference between frozen and unfrozen food is particularly important, since one of the most effective ways of controlling microbial pathogens in food is to keep the food below a temperature at which the pathogen can grow. More radiation is required to kill microbes in frozen food, and the slight heating that results from incidental absorption of radiation by the food has the danger of raising the food to a temperature that would allow pathogenic organisms to grow. Consequently, temperature effects must be carefully monitored in most foods.

Effect of Irradiation on Foodborne Microbial Pathogens

One of the biggest problems with the acceptance of food irradiation in society is that few people are fully aware of how serious the problem of foodborne illness can be. In a way, this reflects well on our existing food safety measures, since it presumably indicates a great deal of consumer confidence in the quality of the food they purchase. Even given the relative safety of the food supply, the harm created by food-related illness is nothing short of staggering. There are five common food borne pathogens which can lead to loss in productivity. They are Salmonella, Escherichia coli, Listeria monocytogenes, Campylobacter jejuni and Vibrio. FDA and USDA have approved doses of 3.0 kGy for poultry, 4.5 kGy for other unfrozen meats, and 7.0 kGy for other frozen meats. Comparing those doses to the D value reveals the percentage of the microbe that will be killed by irradiation at the allowed dose.

- Salmonella: Depending on strain of bacteria and other factors, the D value of Salmonella ranges from 0.4 to 0.8 kGy. At the 3.0 kGy dose approved for poultry, irradiation would kill over 99.9% of the most radiation-resistant strains of Salmonella.
- E. coli: E. coli is even more radiation sensitive that Salmonella; it has a D value ranging from 0.2 to 0.4 kGy. Exposure of beef to a 4.5 kGy dose would reduce the amount of E. coli in the sample by a factor of 100 billion. Considering that E. coli is not ordinarily present in beef, food irradiation could effectively eradicate the problem of E. coli in beef products.
- Listeria: Listeria in beef, pork and lamb has a D value ranging from 0.40 to 0.48 kGy. Using the 4.5 kGy approved dose would reduce the Listeria population by a factor of one billion. Unfortunately, irradiation has not yet been approved for processed meat products such as hot dogs, a common source of Listeria.
- Campylobacter: Campylobacter is one of the more radiation-sensitive bacteria, with the C. jejuni species having a D value ranging from 0.18 to 0.24 kGy. The approved poultry dose of 3.0 kGy would leave only one trillionth of the original Campylobacter population. Considering that Camplylobacter's effects are often chronic and difficult to trace back to food, food irradiation could prevent a great deal of food-related illness that might be undetectable and effectively unavoidable otherwise.
- Radiation-resistant microbe: Most of the discussion thus far has been confined to the most common non-viral pathogens. For the sake of completeness, a few more organisms should be mentioned. C. botulinum, for example, is one of the

more deadly pathogens (although not quite so virulent as Vibrio vulnificus); nearly 8% of hospitalized cases of botulism result in the patient's death. C. botulinum also happens to be quite radiation-resistant in its spore phase, having a D value of between 2 and 4 kGy. Irradiation, at least at conventional doses, would have a limited effect on C. botulinum.

Viruses also play a significant role in food-related illness, and they are much more resistant to radiation that other microbial pathogens. Since people exposed to a virus can often develop immunity to the virus, much of the incidence of food-related viral illness is found in children. Hepatitis A is the most notorious of the viral food-related illnesses. About 5% of all cases of hepatitis A are traced back to food, but since 50% of all hospitalized cases have an indefinite source, the actual number of food-related cases may be significantly higher. Again, irradiation at conventional doses is unlikely to have a significant impact on hepatitis A, and sufficient irradiation could have undesirable effects on the characteristics of infected food (e.g., it might kill raw oysters, reducing shelf life).Another class of disease unlikely to be affected by irradiation is bovine spongiform encepalopathy ("mad cow disease"). Although the cause of the disease is still under study, scientists currently believe the culprit to be prion particles, simple collections of protein lacking DNA. Their simple structure makes them resistant to disruption by radiation.

Chemical Preservation

Chemical food preservatives are substances which, under certain conditions, either delay the growth of microorganisms without necessarily destroying them or prevent deterioration of quality during manufacture and distribution. The former group includes some natural food constituents which, when added to foods, retard or prevent the growth of microorganisms. Sugar is used partly for this purpose in making jams, jellies, and marmalades and in candying fruit. The use of vinegar and salt in pickling and of alcohol in brandying also falls in this category. Some chemicals foreign to foods are added to prevent the growth of microorganisms. The latter group includes some natural food constituents such as ascorbic acid (vitamin C), which is added to frozen peaches to prevent browning, and a long list of chemical compounds foreign to foods and classified as antioxidants, bleaching agents, acidulants, neutralizers, stabilizers, firming agents, and humectants.

Organic Chemical Preservatives

Sodium benzoate and other benzoates are among the principal chemical preservatives. The use of benzoates in certain products in prescribed quantity (usually not exceeding 0.1 percent) is permitted in most countries, some of which require a declaration of its use on the label of the food container. Since free benzoic acid actually is the active agent, benzoates must be used in an acid medium in order to be effective. The ability of cranberries to resist rapid deterioration is attributed to their high benzoic acid content. Benzoic acid is more effective against yeasts than against molds and bacteria. Other organic compounds used as preservatives include vanillic acid esters, monochloroacetic acid, propionates, sorbic acid, dehydroacetic acid, and glycols.

Inorganic Chemical Preservatives

Sulfur dioxide and sulfites are perhaps the most important inorganic chemical preservatives. Sulfites are more effective against molds than against yeasts and are widely used in the preservation of fruits and vegetables. Sulfur compounds are extensively used in wine making and, as in most other instances when this preservative is used, much care has to be exercised to keep the concentrations low in order to avoid undesirable effects on flavour. Oxidizing agents such as nitrates and nitrites are commonly used in the curing of meats.

3

Milk and Milk Products

Technology plays a key role in the processing of milk and manufacturing of milk products. Some of the processes which fall under this category are ultra high temperature processing of milk and milk products, and freezing of desserts. The diverse applications of technology in the production of ice cream, dried milk and cheese have been thoroughly discussed in this chapter.

Ultra-High Temperature Processing Milk and Milk Products

Aseptic processing and packaging (APP) is one of the two major commercial sterilization methods, also called 'in-flow' sterilization. The other commonly used method is 'in-container' sterilization, which generally refers to conventional canning processes. The term 'commercial sterilization' is defined as the absence of any microorganisms capable of reproducing in foods under non-refrigerated conditions of storage and distribution, instead of the absolute absence of every existing form of microbial life.

The first aseptically processed milk in metal cans was developed in Denmark by Jonas Nielsen in 1913, almost 100 years after 1810, when Nicholas Appert discovered the canning process. Consumer-size aseptic packages were first developed in the 1960s in Europe for the milk industry by Tetra Pak. These days milk and some other food products are processed with UHT and aseptic processing in over 60 countries. The market share of UHT milk consumed varies considerably by country: Australia 9%, France 88%, Spain 83%, Germany 63%, Italy 55%, and the United Kingdom 5- 13%.

Basis of UHT Treatment

Heat treatment in the production of long life products is called 'sterilization'. In such processes, the treated product is exposed to such intense heat treatment that the relevant microorganisms and most of the enzymes are inactivated, and the processed product is given excellent keeping qualities and can be stored for several months under ambient conditions. UHT processing uses continuous flow of milk, which renders less chemical change in comparison to retort processing.

In conventional canning process, the product is placed into a container (i.e. metal cans or glass bottle), before sealing and sterilization. However, in APP, the product and the package are continuously sterilized separately. The commercially sterile and cooled product is then filled into the pre-sterilized package in a sterile environment and hermetically sealed to produce a shelf-stable final product with extended shelf life, without the need of refrigerated storage. The processing steps for UHT milk include preheating, homogenizing, holding at preheat temperature, heating to sterilization temperature, aseptic packaging and cooling.

Elements of Aseptic Processing System

A basic aseptic processing system consists of essential elements such as raw product supply tank, pumps with flow control, deaerator, heating section, holding tube, cooling section, back pressure valve, aseptic surge tank, and aseptic packaging system.

Flow Control

Rotary positive displacement pumps are usually the most economical choice for processing homogeneous fluids or fluids with small particles (<3 μm) and the pressure drop in the system is less than 150 psi. Typically, these aseptic pumps have steam tracings around the seals and gaskets to maintain sterility in all critical areas.

Heating and Cooling Sections

There are two main technologies distinguished by the medium used for heating to the UHT, 'direct' and 'indirect' systems. Steam, hot water, and electricity are heating methods for UHT equipment. Heating the product to the holding temperature and subsequently cooling it before packaging are carried out continuously using heat exchangers. These heat exchangers can be classified into two groups:

- Direct heat exchangers, where steam is condensed in the product for heating, and later evaporation removes vapor to cool. Direct heating can take place by the following:
    - Steam injection (steam into product): For homogeneous and high-viscosity products sensitive to shear stress such as creams, desserts and viscous sauces.
    - Steam infusion (product into steam): Also for homogeneous and high-viscosity products sensitive to shear stress such as creams and desserts.
- Indirect heat exchangers, where the product and the heat exchanging fluid are separated by the heating surface. Indirect heating can be done by the following:
    - Plate heat exchangers: For homogeneous and low-viscosity products containing small particles up to 5 μm, such as dairy and juice products. Plate heat exchangers are frequently used for energy regeneration to reduce the cost of heating and cooling.

- Tubular heat exchangers: Not much applied for dairy products.
- Scraped surface heat exchangers: Not applied for dairy products.

The sterilizers utilizing steam or hot water can be subcategorized as direct or indirect heating systems. In the indirect system, the product and heating medium do not have contact, as a barrier (stainless steel) is present. Indirect heating modes include indirect spiral tubes, indirect tubes, indirect plate, scraped surface and electricity. Indirect heating with electricity includes electric elements, conductive heating, and friction.

In 'direct systems', heating to and cooling from the high temperature is very fast due to transfer of the latent heats of condensation and evaporation, respectively, between the steam and liquid milk. Cooling is achieved in a vacuum chamber. Direct heating requires sterilization temperatures 3 to 4 °C higher than indirect heating to achieve an equal sterilization effect because of the greater heat input during the heat-up phase of indirect heating. Systems using a combination of direct and indirect heating are available commercially (e.g., the High-Heat Infusion system of APV and the Tetra Therm Aseptic Plus Two system of Tetra Pak).

Holding Section

After heating, the product is held for a specified time in a holding tube to achieve commercial sterility and then cooled. The divert temperature is the temperature measured at the exit of the holding tube. The holding tube must be maintained under isobaric conditions well above the vapor pressure of the product at the process temperature to prevent flashing or boiling. This is usually accomplished by a back-pressure valve.

Deaerators

Most aseptically processed products must be deaerated prior to thermal processing and packaging. The purpose of deaeration is to remove the entrapped air and minimize oxidative reactions that may reduce the quality of the product during processing and storage. Generally, the deaerator is a vessel maintained at a certain vacuum using a vacuum pump.

Aseptic Surge Tanks

These tanks are used to accumulate the sterile product prior to packaging. They provide flexibility to the process, when the flow rate of sterile product is not compatible with packaging capacity.

Comparison of Indirect Vs. Direct Heating System

A direct heat exchanger is the most rapid method of heating and cooling, presents fewer problems with the formation of scale and burn-on (fouling) and has relatively low initial cost. Indirect heat exchangers give better heat recovery, tend to have a more stable output temperature, and are not prone to contamination from condensable in the steam.

Terminologies Associated with UHT Processing

- Q10 values: It states how many times the speed of reaction increases, if the temperature of the system is raised by 10 °C. The Q10 value for flavor changes, and for most chemical reactions, is around 2 to 3. It implies that if the temperature of a system is raised by 10 °C, the speed of chemical reactions doubles or triples. Q10 values can also be determined for killing of bacterial spores and is normally found in the range of 8 to 30.

- Fo value: It is defined as the number of minutes at 121.1 °C (250 °F) to which the process is equivalent to destroy the heat resistant spore forming microorganism. To obtain commercially sterile milk from good quality raw milk, a Fo value of 5 to 6 is required.

- The F value is referred to as Fo value, when the reference temperature is 121 °C (250 °F) and the z value is 10 °C (18 °F). Different foods have different Fo values, because the thermal resistance of microorganisms can be influenced by a variety of factors.

- B* value: It is based on the assumption that commercial sterility is achieved at 135 °C for 10.1 s with a corresponding z-value of 10.5 °C. This reference process is given a B* value of 1.0, representing a reduction of thermophilic spore count of 109 per unit.

- D-value: It is defined as the required time to decrease microorganism numbers (spore forming) 10-fold at a specified temperature. The D value is the time needed for a microbial population to be reduced by 90%, or 1-log cycle at a constant heating temperature T, and is given by equation:

$$D_T = \frac{\log_{10} No - \log_{10} N}{t}$$

where No is the initial number of microorganisms, N is the number of microorganisms surviving after time t at temperature T.

- Z value: It is the change in temperature from T1 to T2 needed to change the D value by a factor of 10 from DT1 to DT2.

$$Z = \frac{T1 - T2}{\log_{10} DT1 - \log_{10} DT2}$$

Packaging Materials

Aseptic packaging can utilize less expensive and lighter-weight packaging materials such as plastic polymers, in addition to metal or glass containers. However, no single polymer has all the essential properties or meets all the packaging requirements for aseptic packaging. Therefore, a composite multilayered structure is often used to meet package design needs. For example, polyvinylidene chloride (PVDC) and ethylene vinyl alcohol (EVOH) have good barrier properties; high-density polyethylene (HDPE), low-density polyethylene (LDPE) and polypropylene (PP) have good sealing properties; and polyethylene teraphthalate (PET) or nylon has good mechanical and heat resistant properties. Therefore, composites using combinations of these packaging materials can create a heat-resistant rigid container with good sealing strength and gas barrier properties.

There are four types of aseptic packaging: (1) rigid containers, such as metal cans, metal drums, glass bottles, and composites cans; (2) paperboard containers, such as web-fed or roll-fed paper/foil/plastic cartons and preformed/machine-erected cartons; (3) semi-rigid plastic containers, such as web-fed thermoformed or preformed cups, tubs, and trays; and (4) flexible plastic containers, such as blow molded bottle, pouches, sachets, bag-in-box and bag-in-drum. The most widespread consumer package for aseptic products is the paperboard laminated carton. The paper provides mechanical strength, good heat resistance and light protection, but it is coated with polymers due to its sensitivity to moisture. The aluminum foil layer acts as an excellent gas, water, and light barrier, especially when laminated between two plastic layers, and is very thermally stable.

Several kinds of semi-rigid plastic containers are now aseptically produced. HDPE and PP are the two most common thermoplastics used, and pigments are often added to better protect the contents from light. A current trend is the use of PET bottles for aseptically filled beverages, which allow consumers to see the product unlike in traditional aseptic juice box designs.

Sterilization of Aseptic Packages

Sterilization of packaging and food contact surfaces is essential in aseptic systems. A decimal reduction of six is required for sterile, neutral, low-acid products (pH > 4.5). Four main sterilization processes for packaging material are in industrial use, either individually or in combination:

- Mechanical processes: Water rinsing/flushing, air blasting, brushing and ultrasound. Mechanical processes are used primarily for pre-cleaning and reduction of the initial microbial load.
- Heat: Saturated steam, superheated steam, hot air, mixture of hot air and steam, and heat from extrusion during packaging formation. Since high temperature is used, it is not compatible with heat-sensitive plastics.

- Irradiation processes: Ionizing rays, infrared rays and ultraviolet rays. They can be used mostly for heat-sensitive packaging materials or combined with other methods.
- Chemical processes: Hydrogen peroxide (H_2O_2) is the most popular sterilization method for aseptic packaging, because it is relatively fast and efficient.
- Combination of above: Hydrogen peroxide/heat and hydrogen peroxide/UV. Hydrogen peroxide/heat is very popular for the same reasons of hydrogen peroxide alone, except that heat can improve the efficiency further and/or dry the package. A synergistic effect between H_2O_2 and UV is achieved.

Sterilants that are commonly used at industrial level include chlorine, iodine, oxonia, food acids, ozone, H_2O_2, and UV light. The H_2O_2 has proved to be acceptable as package sterilant in the United States. The Food and Drug Administration (FDA) regulations in US specify that a maximum concentration of 35% H_2O_2 for sterilizing food contact surfaces. The residual level of H_2O_2 is regulated with a maximum level of 0.5 ppm. Infrared radiation and vaporized H_2O_2 have been utilized as sterilants for packaging materials. Many other chemicals such as peracetic acid, β-propiolactone, alcohol, chlorine, and its oxide, and ozone have potential for use in sterilizing AP materials.

Aseptic Packaging Systems

The equipment for aseptic packaging can be broadly classified into six basic categories:

- Fill and seal: Aseptic canning systems and systems using glass or preformed plastic containers (thermoformed, injection or blow molded) belong to this group. The preformed containers are sterilized, filled and then sealed in a sterile environment. Examples include Dole, Serac, Remy Metal Box (Freshfill), Gasti, Crosscheck, Hamba, Ampack and Remy.
- Erect, fill and seal: A knocked-down blank of incoming material is erected, sterilized, filled and sealed. Examples include KHS, SIGPKL-Combibloc and Liqui-Pak.
- Form, fill and seal: In this system, a roll of incoming material is sterilized, formed, filled and sealed in an aseptic environment. Typical systems include carton and pouch-forming equipment with examples such as Tetra-Pak, International Paper, Asepak, Impaco, Astepo, DuPont, Prodo-Pak and Thimmonier.
- Thermoform, fill and seal: Roll stock is sterilized or a sterile surface is provided. The material is then heated, thermoformed, filled and sealed in a sterile environment. Examples include Benco, Bosch, Conoffast, Thermoforming USA and IWRA-Hassia.
- Blow mold, fill and seal: The extrudable material is first blow molded. The product is then filled into the package and sealed before the molds are opened, or the

sterile bottle is closed in the mold and delivered to a separate operation, where the sterile container is opened, filled and sealed in an aseptic environment. The process can produce bottles and other semi-rigid containers. Examples include Rommellag, Holopack, Remy, Serac, Bottlepack and Side.

- Bulk packaging and storage systems: Pre-sterilized (or in-line sterilized) bags are aseptically filled and sealed. Drums and totes are first sterilized and then filled and sealed under sterile conditions. The larger size storage tanks are usually sterilized by chemical disinfectants (e.g., iodophores) and/or heat and specially designed valves allow aseptic filling and closing. Examples include bag-in-box systems such as Scholle, FMC FoodTech-FranRica, DuPont Star Asept, Liqui-Box, ELPO and Rapak.

Advantages Gained Using App

The advantages of APP include the following:

- Pumpable foods can be heated rapidly to a very high temperature (e.g. 150 °C), resulting in very short process times (i.e. holding times) to achieve commercial sterility, and then quickly cooled, without the long come-up and cool-down times that is the case with conventional canning process. This method results in preserving higher nutritional quality and has better sensory characteristics.
- A variety of packages can be used in aseptic packaging systems, ranging from traditional glass bottles and aluminum cans to the semi-rigid and flexible polymeric packages or composites of plastic/paperboard/metal; last two being lightweight packages.
- Room temperature can be used throughout storage and distribution, thus reducing energy consumption as compared to a refrigerated supply chain.
- Allows various package shapes and sizes to be used with functionalities such as microwavability, convenience, easy opening and pouring, and great consumer appeal.
- The process is faster and the overall cost is lower, because less energy is needed for sterilization, processing, packaging, storage and transportation.

Microbiology and Chemistry of Aseptic Processing

The extent of thermal treatment needed for commercial sterilization depends on the microorganisms that survive and grow in a product. The most heat-resistant microorganism, known as target microorganism, is chosen to evaluate commercial sterilization conditions. The genera Bacillus and Clostridium are the primary spore-forming spoilage microbes.

D-values or destruction rates for nutrients are usually much greater than those for microorganism. This makes thermal processing feasible by reducing bacteria populations very fast, while retaining food quality. In general, the z values for microorganisms are

5–15 °C, while for nutrients and other quality factors are 20–40 °C or even higher. Thus, HTST and UHT processes are more effective in retaining nutrients and quality than LTLT processing.

Changes in Milk as a Result of UHT Heating

The major challenge in UHT milk production is to impart sufficient heat treatment with minimal flavor change. Direct heating imparts less flavor change but requires more energy in comparison to indirect heating. The residence time is determined by hold tube volume, flow rate and flow rate attributes (viscosity) of specific products. Positive reactions in the hold tube include destruction of bacteria, inactivation of enzymes and hydration of thickeners. Negative reactions include development of off-flavor, initiation of off-color and destruction of vitamins. In general, the heating should be equivalent to a minimum of 9-log reduction of thermophilic spores (referred to as a B* value > 1) and a maximum reduction of 3% in the level of thiamine (sometimes referred to as a Cook value i.e. C* < 1). Heat treatment involves two reactions:

- Type 1 reactions involve the denaturation, degradation, and inactivation of whey proteins, enzymes, and vitamins.
- Type 2 reactions involve the formation of lactulose, hydroxyl methyl furfural (HMF), furosine and others, which are not detected in the raw milk.

Physical Changes

The unwanted physical attributes associated with UHT milk are brown color, sedimentation, protein destabilization, and fat separation. Direct heating after homogenization appears to cause reagglomeration of the small fat globules with the formation of a solid fat layer during storage. To prevent this fat separation, homogenization in direct UHT plants takes place in the downstream position (after the final heating step and vacuum cooling).

Chemical Change

The production of free fatty acids (FFA) during storage is more noticeable in milk with higher fat content and is greater in milk produced in direct rather than indirect systems. Milk proteins change more than any other milk constituent due to UHT processing that contributes to loss of color, flavor, and nutrition, as well as gelation and sedimentation.

Direct heat processing imparts less adverse chemical changes compared to indirect heat processing. In an indirect continuous-flow coiled tube system, the process holding time accounted for >80%, the process heating time <10% and the cooling phase <2% of the accumulated chemical changes. Chemical changes involve flavor, acidity (decreases upon direct UHT process), enzyme inactivation and vitamin decomposition. The volatile sulfydryl compounds are liberated which influences the flavor of the heated milk, deposits form within the heat exchangers, and plasmin, which may cause deterioration of the product, is inhibited. The extent of denaturation of β-lactoglobulin in UHT milk can vary from as low as 35% in direct plants to close to 100% in indirect plants.

Most of CO_2 is lost on heating, with a consequent increase in pH. Milk oxidative rancidity is the reaction of oxygen on milk fat components resulting in short-chain aldehyde and ketone volatiles. Heating has little effect on milk salts with two exceptions, carbonates and calcium phosphates. On heating, soluble calcium phosphate precipitates onto the casein micelles, with a concomitant decrease in the concentration of calcium ions and pH. Some calcium phosphate is rendered insoluble at the high temperatures used in UHT heating and deposits on the surfaces of the heat exchanger (fouling).

Enzyme Inactivation

Enzyme inactivation is a positive chemical change of UHT processing. Thermal inactivation of a transglutaminase (TG) inhibitor provides improved cross-linking of casein micelles, resulting in improved product texture. The milk alkaline proteinase, plasmin, and its inactive precursor, plasminogen, have considerable heat resistance. They are more susceptible to inactivation by indirect UHT processing than by direct heating. Heat-stable proteinases from pseudomonads were reported to retain 20 to 40% of their activity after exposure to UHT conditions of 140 °C for 5 sec.

Destruction of Vitamins

UHT processing reduces B vitamins by 10%, folic acid by 15% and vitamin C by 25%. The fat-soluble vitamins (A, D, E) and some of the water-soluble vitamins (pantothenic acid, nicotinic acid, riboflavin, and biotin) are largely unaffected by UHT treatment, but losses of 20 and 30%, respectively, in thiamine and vitamin B_{12} can occur during UHT treatment. The levels of ascorbic acid and folic acid are markedly reduced in UHT milk containing a significant level of oxygen during UHT processing and storage. The nutritional value of proteins, minerals, and fats is affected minimally by UHT processing and is correlated to the storage temperatures, initial oxygen content and packaging choice.

Heat Stability of Milk

The heat stability of the milk is its ability to undergo high heat treatment without coagulating or gelling. Solutions to improve heat stability include preheating the product in the UHT processor, adjusting pH to the ideal heat stability maximum, and adding phosphate, buttermilk, or phospholipids.

Defects in UHT Milk

Sometimes defects have been encountered in the UHT processed milk, especially during storage. These are discussed herein:

- Cooked and stale flavor: At least two different types of flavors develop in UHT milk. The cooked or heated flavor develops during processing, and the stale flavor develops during storage. Volatile sulfur compounds produced mainly from denatured β-lactoglobulin are responsible for the cooked flavor, which disappears rapidly in the presence of oxygen or oxidizing agents. Aliphatic aldehydes are major contributors to the stale flavor. The heated flavor after UHT processing is due to sulfhydryl (-SH) groups, which oxidize 5 to 10 days after processing. The oxidation then gradually reduces the cooked flavor.

- Chalky or astringent defect: UHT processing causes a slight increase in the size of the casein micelle due to its association with denatured whey proteins and calcium phosphate. This phenomenon can lead to a chalky or astringent defect in UHT milk (especially if heated by steam injection), which can be eliminated by homogenization after the high-heat treatment.
- Sediment formation: Sediment is more prevalent in products that are more severely processed, that have a targeted pH of <6.6 and that have undergone direct instead for indirect UHT processing. Other factors affecting sedimentation include homogenization pressure that is used to control fat separation, time and temperature profile that is used to ensure product sterility.
- Sedimentation in UHT milk is due to destabilization of casein micelles. Sedimentation occurs more readily in concentrated milk than in normal-strength milk.
- Fat separation: Despite homogenization of milk in the UHT process, a layer of fat occasionally develops on the surface of the milk during storage. Less fat separation occurs in milk processed with direct steam injection than with indirect heating, due to the additional homogenization effect of the steam injection.
- Gelation: Gelation during storage is a common problem of UHT milk, which ultimately limits its shelf life. Milk with a high pre-processing microbial count is more susceptible to gel formation than milk with a low count; plasmin has also been implicated. Microorganisms that produce heat-stable enzymes (proteolytic) cause the most serious gelation problems. Longer refrigeration times prior to sterilization allow increased growth of psychrotropic microorganisms and concomitant production of heat-stable enzymes, especially proteinases and lipases. The gelation occurred more readily at room temperatures (20-25 °C) than at low (4 °C) or high (35 to 40 °C) temperatures. Milk sterilized by 'direct heating' gels more rapidly during storage than milk treated by 'indirect' methods. This effect appears to be due to the greater inactivation of the proteinases and greater stabilization of the casein micelle by complexing with denatured whey proteins during the more severe 'indirect heating'. UHT-processed skim milk is more susceptible to gelation than UHT whole milk. This can be attributed to an enhanced action of plasmin and bacterial proteinases in skim milk over whole milk.

Commercial Sterility Testing

An incubation and inspection program is recommended by FDA to verify sterility of aseptically packaged products. Quantitative methods include direct enumeration and viable enumeration. Viable cells are counted using standard plate counts, most probable number, membrane filtration, plate loop methods, or spiral plating. Qualitative methods include measuring metabolic activity or cellular constituents.

Ultra-Pasteurization or Extended Shelf Life (ESL) Processing

The most recent form of heat treatment introduced for milk and other liquid dairy products, such as custard, is ultra-pasteurization or ESL processing, which provides additional shelf life for pasteurized products, which is beneficial to both consumers and retailers. ESL products are products that have been treated in a manner to reduce the microbial count beyond normal pasteurization, packaged under extreme hygienic conditions, and which have a defined prolonged shelf life under refrigeration conditions.

It is a continuous heat treatment of intensity between HTST pasteurization and UHT sterilization, and hence, the properties of ESL milk differ from those of both pasteurized and UHT milk. In particular, ESL milk does not have a strong cooked flavor characteristic of UHT milk. ESL milk can also be produced by microfiltration or centrifugation (bactofugation), with or without a pasteurization step.

A range of different temperature–time combinations have been suggested for producing ESL milk. Generally, temperatures >100 °C for very short times are used, with the most common conditions being 120 °C–130 °C for ~1–4 s. This results in a reasonable extension of shelf life of milk, with little flavor impairment, provided of course that the final heating and initial cooling stages are not excessively long. This is most easily achieved by direct steam heating, either steam injection or steam infusion, e.g., in the APV Pure-Lac™, Instant Infusion, and High-Heat Infusion systems. However, many commercial ESL products are produced on plate and tubular indirect heating plants.

ESL milk is not 'commercially sterile', which means that some (spore-forming) bacteria that can grow at room temperature are not destroyed by the processing conditions. Consequently, all ESL products have to be stored under refrigeration. The shelf life of refrigerated ESL milk is variously cited from 14 to 90 days; a shelf life of up to ~40 days can be achieved with ultraclean packaging, while longer times are possible with aseptic packaging.

Failure Rate in UHT Products

The microbiological failure rate in UHT products is quite low (0.02%). Higher failure rates indicate significant inadequacies in the aseptic packaging step. The spores of some species such as B. stearothermophilus and B. sporothermodurans are extremely heat resistant and can resist UHT treatment conditions. The latter species is of particular concern to the dairy industry as it is mesophilic and hence can grow at the temperature at which UHT milk is normally stored.

Table: Potential failure modes of UHT processed milk.

Type of failure	Reason for failure
Type 1	Raw ingredient, handling, storage, or batching issues.
Type 2	Processor and filler cleaning in place, sanitation, preventive maintenance, and pre-sterilization issues.
Type 3	Thermal process heating cycle including regeneration.
Type 4	The cooling cycle including surge tanks.
Type 5	Sterilization issues with the package.
Type 6	Sterility loss in the aseptic zone or from environmental load.
Type 7	Loss of package integrity.

Other UHT Dairy Products

UHT dairy foods include milk, modified milks, flavored milks, puddings, custards, creams, icecream mixes, and whey-based drinks. Further products introduced were energy and sports drinks, yogurt and sauces. Ultra high temperature non-fat milk, milk, light cream, and 18% cream are used throughout the U.S. by the restaurant and food service industries.

UHT Processed Cheese

According to Joint FAO/WHO Food Standards Program, Processed Cheese means U.H.T., the product defined above and after being processed by heating at 135-145 °C for 5 to 10 seconds or any other combination of time/temperature equivalent. Extension of shelf life of processed cheese is achieved by heating up to 145 °C. For prolonged standing times the UHT can be designed in a double UHT head execution. To reduce setting, product contact parts are PTFE coated. Prior to filling, the product has to be cooled down and deaerated by controlled vacuum, followed by an inclined creaming tank for structure rebuilding and viscosity development under ambient pressure. The Creaming Tank is equipped with a high efficiency creaming element and viscosity monitoring.

MAKLAD UHT cheese plant consists of a pre-heating section (tubular heat exchanger) in which the cold cheese-mix is preheated to about 70 °C by means of ambient vapour, which is generated by vacuum cooler. The preheated cheese-mixture is heated up to 145 °C by an injector by direct steam injection and in one process step. The cheese mixture moves through the heat-retaining zone, where the clostridial spores are inactivated. Post heat-retaining zone, the cheese is continuously cooled to about 90 °C by expansion cooling, under vacuum conditions. After UHT sterilization, the processed cheese mix is extremely liquid. Hence viscosity is increased to the desired level in a creaming tank, by using controlled mechanical agitation and monitoring viscosity by a viscometer. This is finally followed by filling and packing.

UHT Cream

Cream having 15 to 30% milk fat or vegetable fat (imitation creamer) is UHT treated at 142 °C for 3 sec followed by aseptic homogenization at 200 kg/cm^2 pressure at 70 °C. Then it is cooled to less than 10 °C and filled aseptically. The UHT cream may be kept at ambient conditions however, it is best stored at temperature lower than 15 °C. Likewise, UHT whipping dairy cream is also commercially available.

Technology of Dried Milk

Milk is unique in its content of valuable nutrients. Both the chemical composition and the physical properties of the milk powder play an important role in its use with quick and complete reconstitution adding convenience to the products. Every day, millions of liters of milk are processed into dry products. By far the largest part ends up as ordinary skim milk and whole milk powders. These products are market commodities with little requirement for product functionality. Consumer demand for specific product properties, particularly in the food industry, has resulted in the development of many different dry dairy products, ranging from instant whole milk powder to speciality food ingredients. The dairy industry invests heavily in the development and production of such products.

During the last few decades there has been a growing market for powders which are instantly soluble in cold water. Ordinary non-agglomerated powders tend to lump when mixed with water, and if strong mechanical stirring is not applied, it may result in an inhomogeneous mixture which is not attractive to the consumer. The quality requirements to instant milk powders are constantly getting stricter as an increasing number of properties must be optimized simultaneously and controlled within still narrower limits in order to obtain a high quality product with the highest degree of uniformity, fulfilling the requirements set by costumers, organizations and industrial and/or legislative standards.

Instantization

The historic development started with the pioneering research of Mr. David D. Peebles in the beginning of the fifties, and instantized non-fat dry milk was marketed from 1954. Soon it replaced the regular spray dried products on the retail market. Lecithinzing during the drying in the fluid bed is not always applied; it is meant to obtain instant properties.

The principle of instantization is the agglomeration of individual spherical milk particles into clusters and the conversion of lactose from the glass into a microcrystalline form which makes the powder more wettable and less hygroscopic. In the instantization process, the surface of milk powder particles are humidified, or the particles are only partially dried during manufacturing, so that the surface is tacky and partial crystallization of lactose leading to formation of microcrystals occurs before the particles are redried. This produces a clustering of the particles in loose spongy aggregates of low density (known as agglomerates/conglomerates/granulates) which flow freely and disperse readily in cold water, as water penetrates the spongy structure of these aggregates and allow them to sink and disperse.

The principal purpose of agglomeration, also called instantizing, is to improve the rate and completeness of the reconstitutability of dry milk products. Small single particles dissolve instantly in water. Powder consisting of small particles is, however, difficult to disperse. Big particles are easy to disperse in water, but dissolve only partially. Agglomeration is a result of wet and/or semi-dry particle collision with size >125 μm. Control is achieved by returning dry fine powder to the wet spray during different stages of spray drying. Mastering the agglomeration techniques is the art of modern spray drying. Plant operation and economics are other important parameters.

Products

In addition to non-fat dry milk, the instantizing process has been applied to dry whole milk and other milk fat containing dry dairy products with only limited improvement in reconstitutability resulting. The milk fat adversely affects wettability because of its hydrophobic property.

Major Systems

The process may be carried out using either of these two methods:

- Spray Drying Agglomeration.
- Rewet Agglomeration.

Spray Drying Aggomeration

During the spray drying process, the aim is to produce particles with a big surface/mass ratio, i.e. small particles. The reconstitution in water of a powder consisting of small particles is however difficult and requires intensive mixing in order to disperse the powder, before it is totally dissolved. Bigger particles exhibit a better dispersion, but the solubility is negatively affected during the drying operation. By agglomeration both a good dispersion and a complete solution are obtained.

In Forced Secondary agglomeration the small dry particles are introduced into the dryer near the atomizing device, where they will meet and collide with atomized wet droplets thus forming agglomerates consisting of many particles stuck together having a size of 100-500 μ, depending on the parameters selected.

Due to the special air flow pattern in a Multi Stage Dryer, a considerable spontaneous, secondary agglomeration takes place. For production of high quality instant whole or skim milk powder this spontaneous agglomeration suffices, and the fines are just returned to the integrated fluid bed, from where they will get airborne again and reach the atomizing zone again. However, the agglomeration may be further enhanced by forced, primary agglomeration (collision of sprays overlapping each other from different nozzles in a multi-nozzle atomization unit) and/or by returning the fines to the atomization zone (forced, secondary agglomeration). Further flexibility can be gained by designing the atomization unit in a way that allows the distance between the single nozzles or between the nozzles and the fines return tube to be altered. Depending on the atomization device the fines return is designed in different ways:

- Rotary Atomization: In modern dryers fines are therefore introduced from above through the air disperser via four fines pipes situated just above the atomizer cloud. Deflector plates at the end of each fines pipe ensure a correct introduction and distribution of the fines.

- Nozzle Atomization: The fines return is an integral part of the nozzle unit with the fines duct in the centre surrounded by nozzles at the periphery. The fines are introduced tangentially into the fines distribution duct or through a center pipe. The nozzles can be welded to the nozzle rod at a certain angle, so that by turning the nozzle rod around its axis the collision point can be altered.

- Separation: It is the process of separating the part of fines which is entrained in the main drying air leaving the drying chamber. The efficiency of separation is determined by the air flow pattern and air velocities in the drying chamber.

- Attrition: It is defined as the partial break down of agglomerates in fluid beds or powder conveying systems resulting in creation of either fines, smaller agglomerates (abrasion) or of a number of smaller sized particles (fragmentation). Factors affecting the extent of attrition are the jet velocity, determined by the pressure difference across the perforated plate, the fluidization velocity and the actual design of the perforated plate.

It is defined as the separation of fines in fluid beds. The efficiency of classification is mainly determined by the fluidization air velocity and also the fluid bed design. After the final drying, the powder enters the cooling section where the powder is cooled by means of air at ambient temperature followed by cooled, dehumidified air. The powder is finally passed over a sifter where any oversize particles are removed. It is also possible to install a sifter with two nets thus removing any remaining particles/agglomerates of small diameter. Together with the fines, this fraction may be returned to the atomizing device thus producing a powder with a well-defined agglomerate size distribution. The agglomeration is improved by:

- High solids content in the concentrate.
- Bigger quantity of fines returned to the atomizing device.
- Fines introduction closer to the atomizing device.
- Shorter distance from nozzle to fluidized layer in a static fluid bed.
- Higher moisture content from the primary drying stage.
- Bigger primary particles.
- Lower pasteurization temperature of the milk prior to the evaporation.

When leaving the sifter, the powder should not be exposed to strong mechanical conveying, neither by means of air nor by fast moving mechanical screws. However, today's lenient vacuum-low speed air systems are used without too much damage to the agglomerates. The best thing, however, is to install the plant so high that filling into bags or bins is possible by gravity.

Rewet Agglomeration

Since many powders may become instant by the mere agglomeration, many processes have been developed during the past years to agglomerate ordinary powders consisting of single spray particles usually produced in plants with pneumatic conveying system. Each varies in equipment and operation details. The general features in common are:

- Wetting of surface of the particles with steam, atomized water, or a mixture of both.
- Agglomeration, whereby the particles collide due to the turbulence and adhere to each other forming clusters.
- Redrying with hot air.
- Cooling and sizing to eliminate the very large agglomerates and the very small particles.

Wetting

Wetting of the surface of the particles is done with humid air, steam, atomized water either pure or containing milk solids, sugar or other soluble components. The water may further contain additives such as vitamins (water soluble), minerals, colour and surface-active agents. The atomization of the moistening agent can be carried out by means of nozzles or a rotary atomizer. If a rotary atomizer with two feed pipes is used, it is possible to use a combination of steam and water or use two moistening agents, which cannot be mixed for various reasons. If the product is insoluble, an adhesive can be dissolved in the moistening agent. When doing so products otherwise impossible to agglomerate can be agglomerated with a good result.

Agglomeration

Agglomeration, whereby the moist sticky particles collide due to the turbulence and adhere to each other forming agglomerates, is essential for the rewet process. As powders with different compositions do not behave in the same way during the rewetting and agglomeration process, different equipment is needed to obtain an optimal agglomeration. In principle there are two ways of performing the agglomeration: Droplet agglomeration and Surface agglomeration.

Droplet Agglomeration

In the droplet agglomeration process, the powder particles are wetted with droplets of liquid atomized by means of a nozzle or a rotary atomizer while suspended in air. The powder may either be introduced around the rotary atomizer or the nozzle by means of gravity or pressure air conveying, or from below by means of pressure conveying.

The droplet agglomeration process is especially used for powders containing fat such as whole milk powder and powders with a high content of sugar such as cocoa-milk-sugar mixtures. If cold-water instant whole milk powder is produced, a lecithin dosing equipment is installed between two Vibro-Fluidizers. However, the final product quality will never be as good as that produced on a drying plant equipped with fines return, fluid beds and lecithin dosing equipment.

Surface Agglomeration

In the surface agglomeration, either steam or warm moist air with a high relative humidity is used as the moistening agent. The surface of the individual dry particles is wetted due to condensation of the water vapour on the colder particles, whereby the stickiness required for the agglomeration is created. The subsequent agglomeration will take place, if the particles are exposed to sufficient mechanical impact. The warm humid air is usually made by spraying steam into warm air at a given temperature to obtain a relative humidity of 100%.

The surface agglomeration is mainly used for skim milk powder when large agglomerates are aimed at. The final product properties depend to a great extent on the raw material used for the rewetting. The surface agglomeration can also be used for whole milk powder, but the agglomerates get too compact for obtaining a powder with good rehydrating properties.

- Redrying: As the basis powder used for agglomeration is remoistened to obtain the desired surface characteristic for an optimal stickiness, this additional moisture has to be evaporated again in order to reach the specified moisture content. The agglomerates may break down again, if they are exposed to extensive mechanical handling, such as in a pneumatic conveying system. It is therefore necessary to perform the drying in a Vibro-Fluidizer.

- Cooling: Like the redrying, the cooling is performed best in a Vibro-Fluidizer.
- Sizing: Usually, there is a well-defined requirement to the agglomerate size distribution of the final powder. It is therefore necessary to sift the powder. This is done in a sieve with two different net sizes placed above each other. Thus it is possible to remove any agglomerates/lumps considered to be too big in the final product. This oversize fraction may be milled and returned to the process. Powder passing through the upper net may be further fractionated on the lower net into a main fraction and a fines fraction consisting of single particles and agglomerates being too small. This fines fraction is together with the cyclone fraction from the Vibro-Fluidizer recycled back to the process.

Dry Whole Milk

The manufacture of dry whole milk is not much different from non-fat dry milk. Since keeping quality is a restricting factor, care should be taken to obtain the maximum storage life. High quality milk with a low copper and iron content, sanitary plant practices, and a good processing procedure are important influences on keeping quality.

Manufacture

The manufacture of whole milk powder with all necessary stages of manufacture starting from taking milk up to the packaging and storage.

Milk Standardization

The milk fat is standardized in ratio to the solids not fat so the dry product will meet the legal standards. Clarification may be conducted either before or after standardization to remove leucocytes and safeguard against the possibility of extraneous material in the product.

Homogenization

The concentrate is not always homogenized, especially if atomization is done by means of a nozzle, because the fat globules are effectively disrupted in the nozzle. Homogenization of highly concentrated milk considerably increases its viscosity because the transfer of large casein micelles to the fat globules gives the latter such an irregular shape as to increase the effective volume fraction of fat globules plus casein micelles. This increase leads to coarser droplets during atomization, with all drawbacks involved. Consequently, if the concentrated milk is not homogenized, evaporation can continue up to higher dry-matter content.

Homogenization of the whole milk is common if direct reconstitution of the dry whole milk is contemplated. Without homogenization the fat may churn during agitation while combining with water. Another advantage of homogenization is the improvement of keeping quality. The fat globules, although smaller, are more thoroughly recovered in the protein membrane. A pressure of 175 to 250 kg/cm^2 at 62 to 74 °C provides sufficient homogenization.

Storage

Storage (buffering) of the concentrate before atomization is not always applied; it is done mainly to overcome differences in capacity between evaporator and drier. However, the concentrate should not be kept warm for more than a short time to prevent the growth of microorganisms. A refrigerated concentrate generally is too viscous to be atomized readily, and it is therefore heated just prior to atomization because otherwise the viscosity increases again. The heating can at the same time serve to kill bacteria that may have re-contaminated the concentrate.

Preheating

Advantages

- Intense pasteurization is needed to obtain resistance to autoxidation.
- Low-heat treatment minimizes the cooked flavour in the product, but does not develop antioxidants for delay of oxidation, one of the principal factors in keeping quality.
- Usually the primary consideration is given to prolonging the shelf life.
- Preheating also destroys the enzymes. If the lipase enzymes are not destroyed, hydrolytic rancidity will occur in the dry whole milk.
- Preheating must accomplish pasteurization, thus reducing the viable microorganisms.
- A beneficial influence on heat stability of the product may occur from the preheat treatment.
- Heating of the milk also is necessary before it enters the evaporator.

Manufacturing Process

- Direct steam preheating the product in the hot well has been replaced by the indirect method.
- Tubular heaters are currently used in many plants. They eliminate dilution by the steam condensate and steam impurities that may be toxic and/or cause off-flavors.

- Numerous optimum temperature-time conditions have been used in preheating whole milk. Commercial practice frequently employs the range of 82 °C for 15 min to 93 °C for 3 min.
- The heating, after concentration has been suggested as preferable possibly on the basis that a higher percentage of the antioxidants retained. These compounds may be removed along with the moisture during vacuum condensing.
- When sufficient solids concentration as Baum'e degrees has been attained (35 to 45%), product is continuously removed from the evaporator.
- Another variation in the procedure is to preheat skim milk to a high temperature necessary for formation of antioxidants and then to condense the skim milk. Homogenized, pasteurized cream is used to standardize the condensed skim milk to the desired ratio of milk fat to solids not fat. Some operators prefer to homogenize the concentrate after adding the cream.
- The temperature of condensed milk after it is pumped from the evaporator is boosted to 62.8 to 73.9 °C in a heat exchanger prior to the high pressure pump.
- The condensed milk is dried with inlet air at 148 to 232 °C and exit air at 74 to 93 °C, depending upon drier characteristics. To reduce heat damage during dehydration, and yet obtain the desired moisture, a low exhaust air temperature is preferred.
- The dry whole milk should be immediately removed from the hot air stream to maintain better "body characteristics', and keeping quality. The higher the temperature and the longer the time the product is above the melting point of the fat the greater the amount of free fat that results.
- The use of refrigerated air to move the dry whole milk to a cyclone after it leaves the drying chamber is one system of decreasing the temperature. Cooling to room temperature (not below) is preferred. When ambient temperature and humidity are high, the air may be reheated just enough to avoid absorption of moisture by the product due to the dew point. However, other methods of cooling whole milk powder have proved to be satisfactory such as augering across a surface cooled by water or brine in jacketed equipment. The system must be sufficiently protected against the entrance of condensate from the ambient conditions.
- A 12-mesh screen is used for sifting dry whole milk. The product should package immediately or held under vacuum for 7 to 10 days before gas packaging.

Foam Spray Process

Milk is standardized, preheated, and homogenized similarly as for regular spray drying. During evaporation, the solids content may be increased to 50% or above. A gas, preferably nitrogen, is injected into the concentrated milk at 14 (or more) kg/cm^2 greater than the pump pressure. The gas is distributed into the concentrate by means of a mixing device between the high pressure pump and the nozzle. A regulator and needle valve control the flow of gas into the concentrate coming from the high pressure pump at 105 to 126 kg/cm^2. The usage of gas is at the rate of 0.037 m^3 per liter of concentrated whole milk with 50% T.S. Foam spray dried whole milk has improved dispersibility and approximately one-half the bulk density of the regular spray dried product. The occluded air in the particle causes poor sinkability when product is recombined with water.

Drum Process Dry Whole Milk

Only a small amount of drum dried whole milk has been processed in recent years. The poor keeping quality, unsatisfactory reconstitutability, and scorched flavor are considered to be the chief deterrents. However, for uses demanding a high percentage of free fat such as in certain confections, the drum dried product suffices.

Whole milk which has been standardized at the ratio of 3.5 (fat) to 8.75 (solids not fat) will yield a product with 28% fat if dried to 2 % moisture. The standardized milk is pasteurized and may or may not be homogenized. Unhomogenized milk will give a higher free fat in the dried form. The customary homogenization is 175 to 210 kg/cm^2 and the temperature is 62.5 to 76.5 °C. The preheat treatment is regulated for the intended use, but normally is desirable only for dry milk intended for baking products. The severe heat treatment during drum drying negates any attempt to produce a low-heat product. Concentration is not necessary, but will increase the drum drier capacity. Total solids increase should be limited to ~20% concentrate. The procedure and precautions for drum drying of whole milk are very similar to those presented for nonfat dry milk by this process.

Chemical Composition of Human Milk

Chemical composition of human milk differs from that of cow or buffalo milk as shown in tables below:

Table: Chemical composition of human, cow and buffalo milk.

Components, g/100 ml	Human milk	Cow milk	Buffalo milk
Proteins	1.2	3.3	4.2
Fat	3.8	4.1	7.0
Carbohydrate	7.0	4.5	5.1
Ash	0.21	0.72	0.82

Table: Major differences in the makeup of different constituents of human, cow and buffalo milk.

Components	Human milk	Cow milk	Buffalo milk
Proteins:			
• Caseins, % of total protein	40	82.3	75.6
• Whey proteins, % of total protein	60	17.7	24.4
• Immunoglobulins (g/100 ml)	1.0-1.5	0.5	NA
• Lactoferrin (mg/ml)	1.39	0.09	0.32
• Lysozyme (mg/100 ml)	40	0.013	NA
Lipids:			
• Oleic acid (mg/lit)	14.9	10.3	9.5
• Linoleic acid (mg/lit)	3.4	0.8	0.5
• Linolenic acid (mg/lit)	Traces	0.2	0.3
• Arachidonic acid (mg/lit)	0.33	Traces	NA
• Cholesterol (mg/lit)	0.2	0.14	NA
• Lactose	6.9	4.8	5.1
Minerals:			
• Total (%)	0.21	0.68	0.82
• Ca (mg/100 ml)	31	128	210
• P (mg/100 ml)	15	87	128
• Fe (mg/100 ml)	0.15	0.14	6.2
Vitamins & Growth factors:			
• Vit A (mg/100 ml)	0.05	0.03	0.05
• Vit E (mg/100 ml)	0.54	0.07	NA
• Nicotinic acid (mg/100 ml)	0.24	0.08	0.1
• Bifidus factor (units)	40	1	NA

Striking Features of Human Milk

The striking features of human milk when fed to infants are as under:

- Digestion of milk by a baby is speedy: Due to formation of soft curd.
- Quite less load on the kidney: Due to low calcium content.
- Easy digestion and faster absorption of casein: Due to smaller size of casein micelles.

- Proper development of brain constituents: Due to higher lactose content and presence of relatively more amount of PUFA.
- Bactericidal and anti-infective properties: Due to presence of lysozyme, secretory IgA, bifidus factor and lactoferrin.

Human Milk Substitutes

While human milk is superior for the new born infant, human milk substitutes (HMS) play a necessary role in infant nutrition when breast feeding is not possible, desirable or sufficient. Since the advent of first commercial formula developed by scientists of Nestlé's research group, a wide range of infant formulas have been developed by the manufacturers as shown in Table.

Table: Range of infant formulas available in the market.

Type of Formula	Description
Standard Formula	• Use of cow/buffalo milk to replace/supplement human milk. • Adjustments made for protein: fat: carbohydrate ratio. • Vitamins and iron fortified.
Soy-based milk-free	• Use of soy flour or water soluble soy protein isolate.
(SBMS) Formula	• Useful to infants allergic to milk proteins, having lactase deficiency and who show signs of galactosemia.
Specialized/Predigested Formula	• Meant for infants having a variety of nutritional problems due to digestive malfunctions, physiological underdevelopment or gastro intestinal infections.
	• Usually nitrogen is supplied through hydrolyzed proteins, energy through easy to assimilate simple carbohydrates, with minimal load on digestive enzyme.
Humanized/Materialized Formula	• Development of nutritionally superior formulas having varying degree of chemical and biochemical similarity of αs-casein and β-lg from cow/buffalo milk.
	• Adjustment of casein: Whey protein ratio, PUFA and mineral contents.

Approaches to Manufacture Humanized Milk

New generation of nutritionally superior humanized milk formulas has been developed, having varying degree of chemical and biochemical similarity with human milk. There are two types of approaches tried for this:

- Removal/reduction from cow/buffalo milk substances like αs-casein and β-lg through enzyme treatments. However, αs-casein is hydrolyzed in the digestive tract of the infant and thus formula containing hydrolyzed casein may not serve any useful purpose nutritionally.

- Modifications justifiable from nutritional angle-for example adjustment of casein: whey protein ratio in cow/buffalo milk so as to simulate human milk.

Suggestions for Improvements

- Modification of proteins of cow/buffalo milk to make them similar to human milk (whey protein: casein=60:40). This will help in enhancing nitrogen utilization and in minimizing renal osmolar load.
- Modification of triglyceride makeup of cow/buffalo milk fats so as to simulate ratios of short, medium and long chain triglycerides as well as that of saturated to unsaturated fatty acids to that of human milk fat. This will enhance linoleic acid and vitamin E contents.
- Adjustment of proteins: fat: carbohydrate ratio: This will reduce the protein content and increase the carbohydrate content of cow / buffalo milk and make it similar to that of human milk.
- Modification of calcium: Phosphorous ratio as well as reduction in total ash content. This will enhance mineral utilization and minimize renal osmolar loads.
- Enrichment of cow/buffalo milk with factor to support growth of B. bifidum in intestines.
- Enrichment of cow/buffalo milk with lactoferrins, transferrins and immunoglobulins to promote resistance towards enteropathogenic bacteria and to enhance iron absorption.
- Partial substitution of cow/buffalo milk fat with oils rich in PUFA, as well as use of carbohydrate sources such as maltodextrins may help in reducing the cost.

Achievements so Far for Humanization of Buffalo Milk

- αs-casein reduced to human milk level by renneting.
- Total protein concentration reduced to human milk level by proteolysis.
- Calcium level reduced by electrodialysis.
- Curd tension reduced by ion-exchange.
- Level of unsaturated fatty acids improved by corn oil fortification with milk fat.

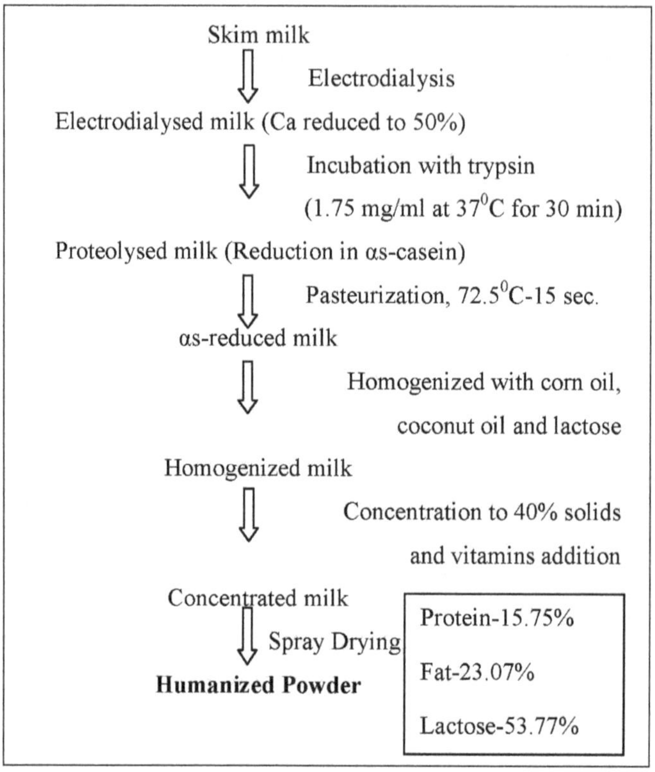

Preparation of humanized buffalo milk powder.

Malted Milk

A British pharmacist James Horlick developed ideas for an improved, wheat and malt-based nutritional supplement for infants. James and his brother William formed a company to manufacture their brand of infant food. The company originally marketed its new product as "Diastoid," but trademarked the name "malted milk" in 1887. Despite its origins as a health food for infants and invalids, malted milk found unexpected markets due to its nutritive value, convenience, digestibility and palatability. It is also appreciated for its lightweight, non-perishable and high-calorie qualities worldwide. Malted milk is a powdered food product made from a mixture of malted barley, wheat flour, and whole milk, which is evaporated until it forms a powder.

Malting

Malting is a process applied to cereal grains, in which the grains are made to germinate by soaking in water and are then quickly halted from germinating further by drying/heating with hot air. Thus, malting is a combination of two processes: the sprouting process and the kilndrying process. Malting grains develops the enzymes that are required to modify the grain's starches into sugars, including monosaccharides (glucose, fructose, etc.) and disaccharides (sucrose, etc.). It also develops other enzymes, such as proteases, which break down the proteins in the grain.

Drying of Malted Milk

The malted milk food is generally prepared by tray-drying method under vacuum which is essentially a batch operation and hence adds to the cost. Spray drying method reduce the cost of product owing to high volume of continuous and automated operation. Thus the production of good quality, low cost malted milk food with improved physical characteristics appears to be better proposition than continuing with the traditional tray-drying method. The method employing spray process is shown in the following flow chart figure.

Butter Milk Powder

Method of Processing

The method of processing of dry buttermilk by either the spray or drum process is similar to that of nonfat dry milk. Care should be exerted to cool the buttermilk to 4 °C immediately upon drainage from the churn. When intended for human consumption, buttermilk should be stored in stainless steel lined tanks. A common practice is to preheat to 32° to 49 °C and separate the buttermilk to reclaim a portion of the fat. The preheating is continued to 85 °C with a 15 min hold. After increasing the total solids by condensing to 16% for drum or ~40 to 45% for spray process, the concentrate is preheated to 71° to 79 °C and dried to a moisture content of 3.0 to 4.0%. After sifting (12-mesh screen), dry buttermilk is usually packaged in kraft paper bags with a plastic liner or in a fiber drum with or without a plastic liner.

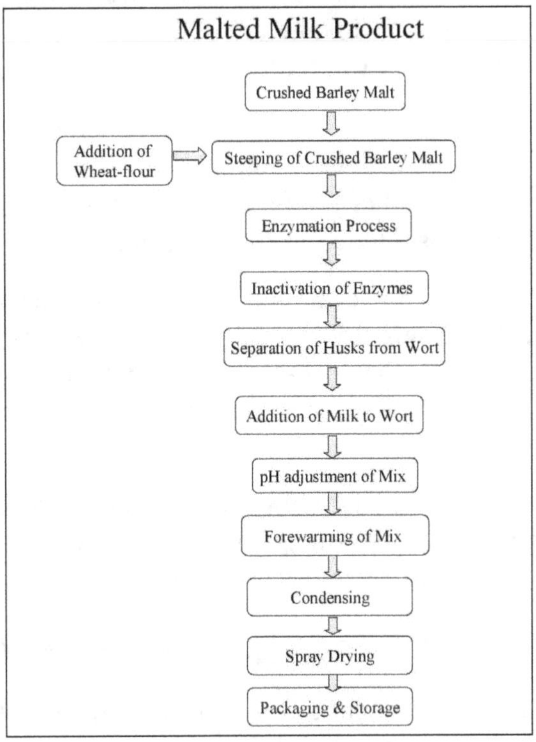

Flow diagram for the manufacture of spray dried malted milk product.

Cream Powder

Cream Powder means the product obtained by partial removal of water from cream obtained from milk of cow and/or buffalo. The fat and/or protein content of the cream may be adjusted by addition and/or withdrawal of milk constituents in such a way as not to alter the whey protein to casein ratio of the milk being adjusted. It shall be of uniform colour and shall have "pleasant taste and flavour free from off flavour and rancidity. It shall also be free from vegetable oil/fat, mineral oil, added flavour and any substance foreign to milk.

Technology of Ice Cream and Frozen Desserts

Ice Cream is a frozen emulsion of air bubbles, ice crystals, milk fat globules, colloidal proteins, and gums suspended within viscous syrup, the continuous phase. Some components of the continuous phase, especially the lactose (milk sugar), are in the supersaturated state. Under certain conditions, lactose may crystallize in the frozen product, causing a sand-like feeling in the mouth. Incorporation of air into ice cream occurs within seconds after entry of the mix into the continuous freezer barrel. As the temperature of the mix drops rapidly in the barrel of the continuous freezer, the air bubbles become entrapped and stabilized within a semi-solid matrix that consists of viscous syrup, ice crystals, and partially churned fat. Properly formed, this matrix can prevent the collapse of air bubbles in the packaged product. Collapse of a significant number of air cells can result in shrinkage within the package and the concurrent loss of quality and acceptability of the product.

Freezing

The quality, palatability and yield of the finished ice cream depend on the freezing of the mix. The freezing process is divided into two main parts:

- The mix, with the proper amount of colour and flavouring material is generally added at the freezer. There it is quickly frozen while being agitated to incorporate air in such a way so as to produce and control formation of small ice crystals that is necessary to produce smoothness in body and texture, palatability and satisfactory overrun in the finished ice cream.
- When ice cream is partially frozen to the proper consistency, it is drawn from the freezer into packages and quickly transferred to cold storage rooms, where the freezing and hardening process is completed without agitation.

Factors that Favor Small Ice Crystal Formation in Ice Cream

- In order to obtain smooth quality that is prized by the consumers, it is imperative to aim for freezing process and thereafter maintaining conditions that will enable prevalence of small ice crystals in ice cream. Such factors that favour small ice crystal formation are:

- Rapid crystallization.
- Agitation while freezing.
- Inoculation with many small crystal centers.
- High viscosity of the medium.

Objectives of Freezing

The objectives of freezing the ice cream mix in order to obtain ice cream are as follows:

- To achieve partial freezing of the product.
- Incorporation of air to give a fine, uniform and stable foam.
- A partial churning of the fat emulsion (in presence of emulsifier).

Freezing in a Batch Ice Cream Freezer

Nowadays, use of batch freezer is confined to plants of moderate production and to supplement continuous freezers in making frozen desserts that are normally not made in large volume such as ices, sherbets and special flavored ice creams. Out of horizontal and vertical batch freezers, the former one is most commonly used since the mix whips up faster. The refrigeration is applied by means of a jacket around the cylindrical freezing chamber forming the jacket. Refrigeration is supplied or shut off by regulating the suction pressure maintained above the liquid ammonia in the chamber. The jacket is covered with an insulating material, which in turn is covered with a sheet metal.

There is provision of a 'dasher-scraper assembly' that consists of dashers, scrapers and expelling device. The dasher can be of several type viz., (i) Blade type, (ii) Squirrel cage type, and (iii) Compound action type. The outermost unit consists of scraper blades and the expelling device, while the innermost unit comprises of dasher or beater. The two units revolve in opposite direction at equal speed (about 140-250 rpm). The speed may vary based on the design and size of the freezer. The speed of the freezer should be such that maximum rates of freezing and whipping are obtained. The scraper blades are hinged so that the centrifugal force and the resistance of the mix will firmly force them against the outside of the freezing chamber. The function of such assembly is to (a) whip the mix, (b) keep the mix scraped free from the freezing surface, (c) circulate the mix in the freezer, from and to end, so that the overrun may be uniform and flavoring materials may be uniformly incorporated, and (d) to hasten the unloading of the freezer.

Direct motor driven freezer is preferred over belt-driven freezer; slippage of belt may occur at later stages of freezing in later case. The temperature of refrigerant should be from -10 to -20 °F, in order to get a rapid formation of ice crystals. Ammonia is the most common refrigerant and has now nearly replaced other types for both batch and continuous ice cream freezers since (i) it is cheaper due to savings in power and equipment, and (ii) it increases the refrigeration capacity. The batch freezer may be of capacity in terms of volume of finished ice cream that can be made per batch, ranging from 20 to 160 quart size (i.e. 22 – 176 lit).

Freezing Operation in a Batch Ice Cream Freezer

- Preparation of the freezer: Clean and dry parts of freezer are assembled using sanitized hands. The freezer chamber should then be sanitized either by use of hot water (180 °F) for about 10-15 min., followed by using cold pasteurized water or use a cold sanitizing (i.e. chlorine or iodophore) solution and draining the said solution.
- Addition of ice cream mix: Ice cream mix at temperatures < 40 °F should be fed to the freezer chamber, after addition of colour and flavor to the mix. The total volume of the mix, flavor and colour should be about half of the total capacity of the freezing chamber. After the mix is filled inside the freezer barrel, the dasher assembly is started immediately followed by turning on of the refrigerant. After about 7 minutes of freezing, the consistency of ice cream can be visually seen through the feed inlet pipe (ice accumulates on the freezer door too). Now the refrigerant is to be shut off and air compressor started and air pressure adjusted to have desired overrun in ice cream, which may be accomplished in about 2-3 minutes. Some ice cream is then drawn out from the freezer outlet, checked for overrun and then directly filled in previously cooled cups or desired pack size. The ice cream should attain stiff consistency when ready to be filled in packages. The temperature of ice cream when drawn from the batch freezer is usually about 24- 26 °F. The ice cream filled in the packages is then transferred immediately to a hardening room for hardening of ice cream.
- Incorporation of particulate inclusions: Special precautions apply when adding acid fruits, nuts, etc. These should be added only after some extent of freezing of ice cream mix has taken place, prior to commencement of whipping.

Cleaning and Sanitizing the Freezer

After the ice cream is emptied from the freezer, the freezer barrel is filled with successive portions of cold and lukewarm (145 °F) water. Add suitable amount of washing powder to the water after rinsing is over. Just a touch of motor switch to give scraper one or two revolutions in the freezer is desirable. Remove the dasher assembly for thorough washing.

Freezing Control Devices in Batch Freezer

- Willman controller: This device is used to indicate 'refrigeration off' point and the 'correct overrun point'. It gives warning when the refrigeration-off point has been reached, on the basis of temperature of ice cream. The 'refrigeration-off' point is determined by judgment for the first few batches and the electric contact point adjusted accordingly. The controller makes use of a timing device which starts when 'refrigeration-off' point is reached and lights a second warning light when a set time has elapsed (indicating point at which overrun will be nearly correct).

- Draw-Rite controller: An ammeter is connected in series with the motor and motor load is marked when the 'refrigeration-off point' is reached (increased ammeter reading). A second point is marked at motor load at the time the desired overrun has been reached (decreased ammeter reading). The refrigerant is shut off when the first marked point is reached and mix drawn from the freezer, when the second marked point is reached.

Freezing in a Continuous Ice Cream Freezer

Principle of Continuous Freezing of Mix

In this system, the mix and air are fed into the ice cream freezer by arranging two feed pumps in series, with an air inlet valve between them. The first pump regulates the rate of mix flow while the second one operates at a higher speed than the first. This together with regulation of inlet valve, controls the quantity of air drawn into the mix and provides the pressure for forcing the mixture through the freezer barrel.

Continuous Ice Cream Freezing Operation

A typical scraped surface freezer (SSF) consists of a cylindrical barrel with a refrigerant, such as vaporizing ammonia or Freon, surrounding it. Inside the barrel is a rotating shaft, or dasher, with scraper blades attached to it. Typically, the barrel and dasher assembly are made of stainless steel. Ice cream may be frozen either in a batch or continuous process. A batch process may take about 10 to 20 min to freeze. A continuous freezer, common to large operations, has a residence time on the order of 1 to 2 min. The refrigerant temperature is about −30 °C.

Description of Continuous Ice Cream Freezer

There is provision of a freezer barrel. Such barrel is surrounded by a concentric outer cylinder carrying refrigerant of sufficient capacity to freeze the mix flowing through the barrel to a plastic mass. The mix is frozen by the time the mix passes through the length of the barrel. The mutator shaft rotates at several hundred revolutions per minute within the barrel. The shaft carries knife blades so arranged that they continuously scrape the ice cream from the barrel surface, as it freezes there.

Since the space between the mutator and the freezer barrel is less, the product passes through the barrel in a few seconds (5-15 sec) giving rise to 'instant' freezing. A 'solid mutator' is used for production of harder, extruded ice cream, while for soft ice cream an 'open mutator' is available. Hollow mutator overcomes the problem of sporadic ice accumulation on mutator resulting in high back pressure on the pumps. Moreover, they provide greater flexibility in the freezing and whipping conditions. The clearance between the mutator and the wall of the freezer tube is only ~ 5/16". As a result, the mix is in the freezer for only 5-15 sec. resulting in 'Instant' freezing.

Dashers are made in a variety of displacements, the percentage of the freezer barrel volume occupied by the dasher. Solid dashers have high displacement, typically about 80%, whereas open dashers have low displacement and often include beaters. Pressure is maintained inside the barrel by a back pressure valve at the outlet of the freezer. Earlier models used spring loaded valves, in later machines, it is controlled pneumatically. Maintenance of correct back pressure helps in proper incorporation of air in small uniform cells throughout the mass of product during freezing.

Working of a Continuous Ice Cream Freezer

The mix and air are pumped into the machine by means of a two-stage pump (two separate pumps). The first stage pumps just the mix, while the second stage pumping more rapidly (twice the capacity of first pump), creates suction between these two stages and permits introduction of mix in controlled amounts. The second stage also provides the pressure which propels the product through the freezing chamber and through the discharge pipes. The air supply is drawn in through the air valve, which can be adjusted (20-30 psi pressure) so as to obtain the desired amount of overrun. At the freezer outlet a spring loaded valve/pneumatic valve assures back pressure on the freezing chamber so that air may be properly incorporated in ice cream.

To produce ice cream in the typical commercial process, the ingredients are first mixed together, then pasteurized, homogenized, cooled, and aged for at least 4 h at a temperature of about 4 °C. Flavoring is then added before the mix is pumped into the scraped-surface freezer (SSF), where about 50% of the water is frozen and air is incorporated into the product. Upon exiting the SSF, inclusions and variegates may be added, and the soft product, at about −5 to −6 °C, is filled into its retail container. In case, nut or fruit variety is to be made, the fruit or nut inclusion particulates can be added into the exiting partly frozen ice cream through use of a 'Fruit feeder machine'. The fruit feeder is connected to the outlet of the freezer.

The ice cream is sent to harden until the core reaches a specified temperature, usually about −18 °C. As the temperature drops and more ice crystallizes, the remaining water contains more and more sugar, which depresses the freezing point so that in the final, hardened product, about 75-90% of the water is frozen. The finished product is stored between −18 and −30 °C, depending on the plant, distributed, and sold. In the supermarket, temperatures in certain types of freezer cabinets (particularly open-faced cabinets) can reach −9 °C, and in frost-free consumer freezers, temperatures can vary quite widely during the frost-free cycle.

Changes that Occur in Ice Cream Mix During Freezing

Initial freezing is a dynamic process: the mix is frozen while being agitated, which whips in the air, destabilizes the fat, and scrapes ice into the bulk fluid. Cold ice cream mix enters the freezer, with the refrigerant absorbing the heat in the mix until the super-cooling at the wall is great enough to initiate ice nucleation. As ice crystals form at the wall, the dasher blades scrape the surface layer from the wall about every 0.075 s, assuming 4 blades with a dasher speed of 200 rpm, and propel the ice layer, at a temperature of −20 to −25 °C, into the bulk of the freezing ice cream mix. When the ice cream exits the freezer, it is usually around −5 to −6 °C, and close to 50% of the water is frozen. The average ice crystal size exiting an ice cream freezer is 20 to 30 μm.

Ice crystal size could increase with dasher speed because of the extra mechanical energy and frictional heat being put into the system via the rotation of the dasher, which accelerates recrystallization. Conversely, increasing dasher speed could help reduce ice crystal size by scraping off a thinner ice layer. Hence it is reported that an optimum balance between rotation rate and heat generation is required, which would depend on the individual freezer and ice cream mix composition to produce the smallest ice crystals. The operator's principal work is to:

- Regulate the amount of air being introduced into the mix to give the desired overrun.
- Regulate the temperature of refrigerant in the freezing chamber.

Recirculation Continuous Ice Cream Freezer

One potential modification to the typical ice cream freezing process is to recirculate a portion of the product stream back to the freezer inlet stream. The benefit of this practice is mainly for startup, where recirculation minimizes production of unused unfrozen product. The recirculation stream brings existing ice crystals into the mix where they can serve as seed crystals and can continue to grow in size. This results in a wider residence time distribution with a higher average residence time and, thus, a larger, wider crystal size distribution. Recirculation has a beneficial effect on the air cell size, however, because of the extra whipping (through increased residence time), breaks the air cells into smaller bubbles.

Low-Temperature Continuous Freezers

In some models, there is a second freezer barrel so that ice cream in fact receives a second freezing – device that increases the capacity of machine. By rotating mutator at high speed in first barrel when ice cream is in early stages of freezing and therefore of lower viscosity. The mutator is at a lower speed in the second barrel, when freezing process is near completion and the product viscosity is high. In low temperature continuous freezer, ice cream can be drawn at - 15 °C. This type of freezer is useful for stiff extrusions or for filing bulk cans (e.g. 2 gallon cans).

A new technology for reducing draw temperature and eliminating the need for hardening is cold extrusion. The process reportedly improves both the sensory and shelf-life properties of the ice cream. The low temperature extruder is placed after the normal SSF and can bring the draw temperature down to −15 to −18 °C. The extruder may be in either a single-screw configuration, which churns the ice cream around the screw, or a twin-screw configuration, which kneads and churns the ice cream between two parallel screws. Both operate at a rotational rate of about 15 rpm. In the extruder, high homogeneous local shear breaks up the fat globules, ice crystals, and air cells into smaller units which improve the flow ability of ice cream. Compared to conventional ice cream freezing and hardening, including a twin-screw cold extruder reduces ice crystal and air cell size by a factor of 2-3. The smaller fat globule aggregates resulting from the high shear in the cold extruder impart a creamy texture to the ice cream, without a buttery defect, making the technology useful for low-fat formulations. Such ice cream is more resistant to adverse handling conditions than regular frozen ice cream.

CIP of Continuous Freezer

- Rinse with water (≤ 100 °F) until the rinse water runs clear.
- Flush for 20-30 min with 150-160 °F water containing 0.5-0.7% alkali strength detergent.
- Rinse until the equipment is cooled.
- When using acid cleaner, circulate cleaning solution containing sufficient acid (phosphoric and hydroxyacetic acid) to give 0.5-0.6% acidity, at 150-160 °F and 5-7.5 ft./sec velocity for 20-30 min.
- Drain and rinse with water at 145 °F for 5-7 min.
- The freezer is made to run 10 sec for every 10 min. of cleaning operation.

Structure of Frozen Ice Cream

The dispersion and emulsion consist primarily of a freeze-concentrated aqueous serum phase containing sugar and the dry matter contents surrounding dispersed ice crystals and fat globules. Ice crystals range in size from ~ 1 to > 150 µm in diameter, with an average size of 35 µm. Fat globules are ≤ 2 µm in diameter. The foam is formed by pockets of air (~ 20 to 50 µm diameter) dispersed throughout the emulsion and is supported by partially coalesced fat globules.

In 'Low-temperature freezing', the ice crystal size may be reduced as much as 40%

e.g. from 45-55 um to 18-22 μm. The air cell size seems to remain about the same or become a little larger, but the number of air cells increases slightly in number. The air cell wall thickness is just 50-100 μm as against 100-150 μm in ice cream frozen in normal continuous freezer. Ice crystal size is also important to ice cream shelf life. As the ice cream sits in storage, the ice crystals continually grow by recrystallization. The temperature fluctuations seen in a frost-free home freezer serve to accelerate this process.

Hardening, Storage, Distribution and Defects

After ice cream is drawn from the freezer, it is filled into containers to be placed for hardening; freezing process is continued without agitation until the temperature of ice cream reaches 0 °F or lower, preferably -15 °F (-26.1 °C). Quick hardening is desirable since slow hardening favours formation of large ice crystals and corresponding coarseness of texture in ice cream.

Hardening Time

The time necessary for the temperature of the center of the package to drop down to 0 °F is known as 'Hardening time'. When hardening tunnels are used, the rate of hardening is several times faster.

Table: Hardening time as affected by type of hardening method and the pack size.

Type of hardener	Package size	Hardening time
Still air hardening room	¼ pint	30 min
Still air hardening room	5 gallon	24 h
Contact Plate hardeners	5 lit	2-3 h
Cryogenic hardening	5 lit	1 h

Factors Affecting Hardening Time

Some of the important factors that influences hardening time are:

- Size, shape, colour and type of package.
- Air circulation in the hardening room.
- Temperature of the air.
- Location of ice cream pack in the section of the hardening room.
- Temperature of ice cream drawn from the freezer.
- Composition of the ice cream mix.

- Percent overrun in ice cream.
- Temperature of the surroundings.

Types of Hardening Units

Traditional method employed Hardening rooms. Hardening cabinet resembles the retail ice cream cabinet and is refrigerated by mechanical refrigeration. These are usually operated at a temperature between -10° and -15 °F and are most economical for a limited volume of business. The hardening rooms are classified into (a) Still air hardening rooms, (b) Gravity air type hardening rooms, and (c) Forced-air type of hardening rooms. Quicker hardening methods are being adopted. These include: Blast tunnel hardeners, Plate or Contact hardeners and Cryogenic hardening using liquid nitrogen.

Some manufacturers of larger volumes use hardening tunnels that produce an air blast at -30 to -50 °F for rapid hardening. These may or may not contain a conveyor belts. The conveying systems have been expanded to include the wide flat belt, fixed tray, suspended free tray, and multi-shelf carrier types of conveyors. The zone hardening tunnel and the ceiling conveyor systems are other types of hardeners.

- Blast tunnel hardeners: Ice cream is hardened in a hardening tunnel, an enclosed chamber into which the ice cream passes on a conveyor belt from the factory freezer. Inside, cold air (typically – 30 °C to – 45 °C) is blown over the ice cream. The lower the air temperature and the faster the air flow, the faster heat is removed from the ice cream. Air turbulence also increases the rate of heat transfer.
- Contact Plate hardener: It is a device used to accomplish rapid hardening of ice cream. It involves an arrangement in which two opposing sides of ice cream packages are placed for a controlled time of contact with two parallel hollow plates, through which cold brine or expanding refrigerant passes. This leads to quite rapid hardening of ice cream in packages.
- Cryogenic hardening: This involves hardening of ice cream using liquid nitrogen (i.e. passing ice cream through Cryo's LIN Tunnel freezer at Gulf) which leads to quick hardening of ice cream in bulk in less than 1 h compared to more than 24 h in conventional hardening. The nitrogen remains liquid at temperatures ranging from -346 to -320 °F.

Storage of Ice Cream

After ice cream is hardened, it may be immediately marketed, or it may be stored for a week or two at the most. The manufacturers plan on a maximum of 5 days between freezing and marketing. At least 12 h of this is required for hardening ice cream which frequently remains in the hardening room until marketed. Since hardened ice cream can be stored satisfactorily at a slightly higher temperature than are required for hardening, it is sometimes more economical to use special storage rooms. The temperature should be maintained uniformly at a point between -10 and 0 oF. The packages should be piled very closely to delay any changes in the temperature of the ice cream.

Shipping of Ice Cream

During marketing, the manufacturer ships it to the retailer under refrigeration at the same temperature as is maintained in the Retailer's cabinet. Dry ice (solid CO_2) may be used for package deliveries. The dry ice is sawed into pieces of appropriate size, which is wrapped in paper to delay rapid evaporation. Then it is placed around the package of ice cream inside an insulated packer, or in single-service type packer. The latter are usually cardboard boxes insulated with corrugated cardboard and are used especially for carryout packages.

Defects in Ice Cream

Ice cream to be marketed has to possess all of the ideal characteristics associated with the variety of ice cream produced. However, defects may arise in ice cream as a result of use of subgrade raw materials, improper balancing of the ice cream mix, faulty processing, freezing and whipping, hardening and 'heat shock cycle' (temperature fluctuation in the Retailer's cabinet) during storage of ice cream. Any deviation from the 'ideal characteristics' of ice cream indicates that the product may be 'defective'.

Table: Ideal characteristics of ice cream.

Attributes	Ideal characteristics of ice cream
Colour and appearance	Delicate, appropriate colours, uniformity of colour should be there. The ideal colour is characteristic of the flavouring used, true in shade, neither too pale nor too intense. Not only uniformity of layers, but also clearness of object outlines should be evident in molded ice creams.
Flavour	The flavor should be typical, pleasing and desirable. The flavor of flavouring added to ice cream mix should be perceptible and should not give harsh note.
Body and texture	Ice cream should be smooth, the solid particles (ice crystals, lactose crystals, etc.) being too small to be detected in the mouth.
Melting quality I	Ice cream should slowly melt down to produce a liquid of similar appearance to the mix from which it was prepared. Ice cream should melt down to a non-foamy liquid to present most attractive appearance.
Package	The package should be clean and attractive in appearance; the product inside the package should be completely covered. If cans are used, they should not have rusted.

Defects may result from faults in flavor, body and texture, melting characteristics, colour and package, bacterial content or even composition.

Defects in Flavour

The ideal flavor should be typical, pleasing and desirable. The types of flavor defects that can probably occur in ice cream are listed in Table. The sources of flavor defects maybe:

- Dairy products of poor quality.
- Sweetness: Excess or deficient.
- Flavouring: Excess, deficient or atypical.
- Blend: May not be pleasing.
- Serving conditions: Too hard or too soft product.

Table: Flavour defects in ice cream.

Type of flavor defect	Characteristics	Causative factor
Harsh or low flavour	Excess, unappealing flavor or flavor not perceived aptly.	Flavouring ingredient is harsh or not having desired strength of flavour.
Unnatural flavour Imparted by dairy products		
Imparted by dairy products		
Cooked flavour	Higher intensity of cooked (heated) flavour.	Use of heat treated dairy products viz, evaporated milk, khoa, etc.
Feed flavour	Flavour of feed and weed consumed by milch animals.	Milked from milch animal fed on weed and pungent feed.
Imparted by tainted dairy products		
High acid	Sour taste.	Sour milk ingredients used.
Old ingredient	Yeasty, cheesy, musty flavor or bitter taste.	Ingredients used were not fresh and had undergone storage changes.
Rancid/oxidized	Bitter taste.	Ingredients used were oxidized.
Salty		
Flavour defect by chemical changes		
Stale and unclean	Flavour of flavouring not quite perceptible.	Surrounding of ice cream package not clean.
Lacks fine flavour	Lacks desired flavor quality.	Flavouring used is unnatural or interfered by other reactions.
Acid flavour	Sour taste.	Milk ingredients used were acidic.
Bitter flavour	Bitter flavour.	Rancidity of fat source.
Flat flavour	Flat flavour.	Storage induced flat flavour.
Metallic/oxidized flavour	Rancid or metal like taste.	Exposing fat to sunlight or diffused light.

Defects in Body and Texture

The causative factors for body and texture defects that may be encountered in ice cream are:

- Improper composition of mix.
- Improper processing methods.
- Improper storage conditions.

Body Defects

- Weak: Ice cream lacks firmness or chewiness. When tasted, ice cream is watery and body is fairly light.
- Causative factor:
  - Low TS content coupled with insufficient stabilization.
  - Withdrawing ice cream from freezer before it has properly frozen.
  - Protein destabilized in ingredients.
- Crumbly: Lacks cohesion and pulls or breaks apart very easily when served.
- Causative factor:
  - Use of too little stabilizer.
  - Low homogenization pressure.
  - Too low sugar or TS content.
  - Too high overrun, with large air cells.
- Heavy and soggy: Soggy body is dense, may be somewhat 'wet' in appearance, tastes much colder than normal and is unattractive in appearance.
- Causative factor:
  - Low overrun, especially when TS is high.
  - Excess concentration of stabilizer.
- Gummy/chewy: Ice cream dishes as a doughy or pasty mass. Ice cream needs to be chewed in mouth.
- Causative factor:
  - Excessive stabilization, especially with gelatin or gum.
  - Excess homogenization pressure.
  - Too high TS content.

- Pasty, sticky or gluey: The ice cream tends to stick to the spoon and trails off when spoon is withdrawn from package.
- Causative factor:
    - Use of syrups or certain types of gum (pectin, oatgum, etc).
    - Use of too much egg yolk solids.
    - Use of superheated condensed milk and not reducing stabilizer content.
- Foamy.
- Dry: Ice cream tastes very dry and is not very appetizing.
- Causative factor:
    - Excess amount of emulsifier used.
    - Excess homogenization pressure.
    - Excess use of certain types of vegetable stabilizers.
    - Addition of dry milk solids at the freezer.

Texture Defects

- Coarse/Icy: Ice crystals are large or not uniform in size; even air cells are too large.
- Causative factor:
    - Low TS content of mix.
    - Slow freezing (dull freezer blades), slow freezing during hardening, fluctuating storage temperature.
- Snowy/Flaky/Fluffy: Amount of air incorporated in mix is too great and present as large air cells which makes texture of ice cream light and snowy; it may appear flaky too.
- Causative factor:
    - Low TS and high overrun.
    - Freezing ice cream too soft in the freezer.
    - Use of low stabilizer content.
- Buttery: Lumps of butterfat, easily detected in mouth while consuming ice cream. As a result of churning of fat, especially during freezing process.
- Causative factor:
    - Incomplete homogenization.

  - Using too high (> 16%) fat content as in Superpremium ice cream.
  - Use of too much white butter in the mix
  - Freezing mix too stiff in the freezer.
- Sandy: Roughness likes sand in melted ice cream that is noted not only when rubbed against roof of mouth but also when chewed. It is due to fairly large (> 25 µm) lactose crystals which are slow to dissolve.
- Causative factor:
  - High lactose content (i.e. > 8.5%) in relation to amount of water present in mix.
  - High and fluctuating temperature in retail cabinet.
  - Low viscosity of unfrozen liquid phase.
  - Substances present in ice cream (nut piece, cocoa powder) that initiates crystal formation.

Melting Quality Defects

- Foamy meltdown: Melted appearance of ice cream is distinctly foamy.
- Causative factor:
  - Incorporation of too much air in ice cream, especially due to large air cells.
  - Large amount of egg solids.
  - High overrun through use of Nacaseinate or some stabilizers.
- Curdy meltdown/Scummy appearance on melting: Appearance of slightly greenish serum that separated out from the more solid ice cream during melting.
- Causative factor:
  - Use of stabilizers that destabilize milk protein (i.e. locust bean gum, guar gum, CMC).
  - Excess homogenization pressure Melting and refreezing in the deep freezer.
- Whey leakage: Whey separates out from ice cream upon melting.
- Causative factor:
  - Mix is of poor quality.
  - Mix is improperly balanced or stabilized.
- Slow meltdown: When ice cream melts slowly, retains its form after melting, or breaks apart in chunks.

- Causative factor:
  - Improper processing of mix.
  - Homogenizing at too low temperature of mix.
  - Use of excess quantity of stabilizers.
- Very high melting resistance: Ice cream does not melt within 1 h of serving or more.
- Causative factor:
  - High fat content in ice cream.
  - Use of calcium neutralizers Use of certain type of stabilizers.
  - Drawing ice cream at too low a temperature from freezer.

Colour and Package Defects

- Unnatural colour: Insufficient (pale) color, excessive (intense) colour or colour that are not characteristic (true in shade corresponding to the flavor used).
- Causative factor:
  - Carelessness in adding colour.
  - Improper use of colour.
  - Use of foreign materials.
- Uneven colour: Colour not uniformly distributed throughout the matrix of ice cream.
- Causative factor:
  - Colour not properly added.
  - Not exercised care when changing flavours in ice cream.

Miscellaneous Defects

- Shrinkage: The volume of ice cream shrinks (may lose its shape too), leaving a space either at the top or side of the package, which appears as 'not full'. It may develop within 3 or 4 days or even require 3-4 months to manifest itself.
- Causative factor:
  - Use of untreated paper containers; loss in moisture from ice cream during storage.

- Temperature fluctuation during storage Excessive overrun in ice cream.

Hence, it is imperative to take utmost care starting from ingredient selection for ice cream mix preparation up to storage and distribution, so that consumers can enjoy the frozen delicacy in its ideal condition as manufactured by the ice cream industry.

Technology of Cheese Manufacture

Cheese is a highly nutritious food. It is convenient and versatile and offers a diversity of flavours and textures. The global sale value of cheese represents about 30% of total dairy products sales with a forecast of a 9.8% growth in cheese sales between 2003 and 2007. The growth in the cheese sector is due mostly to convenience packaging, increased use of cheese in the food service sector, and growth in speciality and exotic cheeses. The USA, the UK, France and Germany experienced more than 9% growth in the cheese market during the five-year period.

Table: Milk and cheese production in selected countries (1000 metric tons).

	USA		EU		Russia		New Zealand	
	Milk	Cheese	Milk	Cheese	Milk	Cheese	Milk	Cheese
2000	76004	3746	114900	5437	31900	220	12236	297
2001	75067	3747	114856	5420	33000	260	13162	281
2002	77002	3900	115130	5470	33500	340	13925	312
2003	77293	3975	115150	5500	33150	320	14346	272

The manufacture of cheese is a form of milk preservation as milk is highly perishable. All cheeses, whether rennet or acid set, can be classified as soft, semisoft (semihard), hard, or very hard, depending on their moisture content.

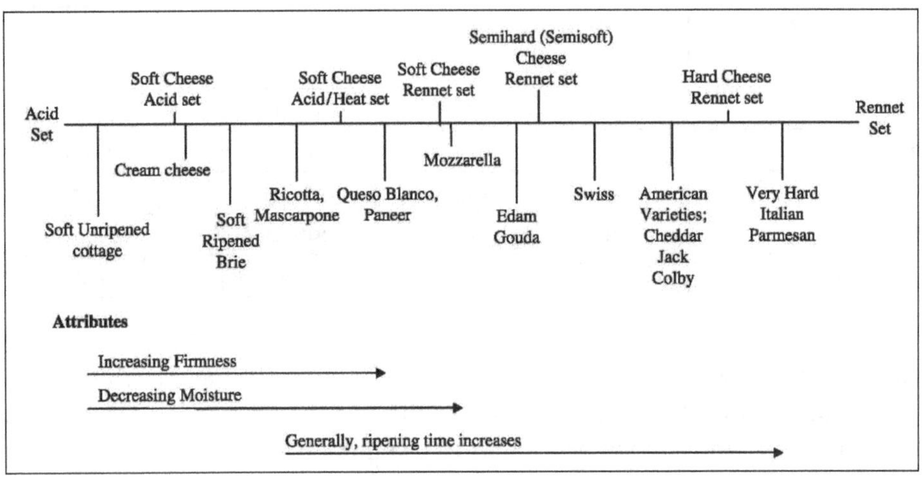

The cheese spectrum.

Although this classification is arbitrary and utilitarian, it helps to systematically group together cheeses that are alike in certain basic features or characteristics (e.g. moisture content), as moisture determines the body, consistency or compactness of cheese. Therefore, the term soft cheese is used to describe cheese that is soft to touch or to pressure applied between fingers.

Table: Growth in the cheese market in selected countries.

	US$ billion (2002)	% Growth in sales (1998–2002)	Reason for expansion
US	13.3	29.7	Convenient packaging, increased varieties.
UK	2.6	9.4	Ingredient use of cheese, convenient packaging.
France	6.9	12	Pre-packaged cheese, foreign cheese.
Germany	5.64	13.8	Exotic and foreign cheeses (e.g. feta, mozzarella).

Conversely, the terms hard cheese (e.g. Cheddar) or very hard cheese (e.g. Parmesan) refer to cheeses that are firm or very firm, respectively, and require some form of pressure to break apart. Classification of cheese based on moisture content is given in Table. Semisoft, hard and very hard cheeses have upper limits for moisture content and lower limits for fat content– usually expressed as fat in dry matter (FDM). The US Code of Federal Regulations specify that soft cheeses must contain a minimum of 50% FDM, but does not specify a maximum moisture content However, there is a practical limit for moisture of about 80%, above which the product is no longer a semisolid mass but is liquid.

Table: Global value sales of selected dairy products by sector (US$ million, rsp).

	1998	2003	2007	% Growth 2003–2007
Dairy products totals	218881.5	222971.3	247944.2	11.2
Milk	76080.0	73067.7	78242.8	1.4
Cheese	66584.5	67810.0	74439.1	9.8
Yoghurt	24363.0	23334.4	34002.1	20.0
Cream	8863.7	8934.4	9769.9	13.7
Flavoured milk	7424.2	8367.1	9963.1	19.1
Milk powder	6539.3	6003.1	7309.1	21.8

The method used to clot milk for cheese making influences the overall structure, characteristics and firmness of the cheese. The two basic methods for clotting milk for cheese manufacture are by rennet or acid, leading to the respective terms, rennet- or acid-coagulated cheeses. In general, acid-coagulated cheeses are soft whereas rennet-set cheeses are firm.

Milk Quality and Pre-treatment Before Cheesemaking

Hydrogen Peroxide/Catalase Treatment of Milk

The starting material for good quality cheese is good quality, clean-tasting milk that is low in somatic cell count, free of antibiotics and has a relatively low microbial count. The composition of milk, which is influenced by the breed of cow, season, stage of lactation, disease and genetics, affects the yield, quality and functional characteristics of cheese. Depending on the microbial quality of the raw milk and its source, and to ensure consistent cheese quality, raw milk may be treated with hydrogen peroxide (H_2O_2). This treatment is not currently used in technologically advanced countries with hygienic milking conditions. Hydrogen peroxide is both a bactericidal and a bacteriostatic agent. It is normally used at 0.07–0.1% for a maximum of 40 min at 50–54 °C. H_2O_2–catalase treatment destroyed almost all the coliforms in milk. The role of catalase is to decompose the toxic effect of residual H_2O_2 and the amount of catalase added.

Bactofugation

The introduction of the bactofuge has helped to control the quality of milk in regions of the world where the bacteriological quality of the milk is poor. were the first to demonstrate the effect of bactofugation on the control of microbiological quality of cheese milk. They showed that bactofugation of milk decreased bacterial numbers by 95.3%. The bactofuge is a high-speed centrifuge specially designed for removing bacteria and bacterial spores from milk at high temperatures. A double bactofuge treatment at 73 °C results in at least three log reductions of bacteria and bacterial spores. Bactofugation reduces the initial volume of milk by 2–3% and the milk protein content by ~7% results in cheese yield losses of ~6% because of the loss of milk solids in the sludge. The sludge is ultrahigh temperature (UHT)-treated at 130–140 °C for a few seconds and then added back to the cheese milk.

Table: Classification of cheese based on moisture content.

Consistency	Maximum moisture (%)	Minimum fat-in-dry matter (%)
Hard grating	34	32
Hard	39	50
Semisoft	50(>39)	50
Semisoft, part skim	50	45(<50)
Soft	Not specified	50

Chemical Treatment

Even when milk is subjected to bactofugation treatment, a certain amount of nitrate (~2.5 g/ 100 kg milk) is desirable to inhibit spore germination in countries that allow the use of nitrates in cheese making. Nitrates are not allowed in cheese making in the USA. Alternately, lysozyme (100–250 mg/100 mL) may be added to the milk.

Lysozyme associates with casein micelles and hydrolyses peptidoglycan bonds of mostly Gram-positive bacteria (e.g. Clostridium tyrobutyricum) that tend to grow, when present, in varieties such as Swiss cheese and causes late-gas blowing. Although lysozyme occurs in bovine milk at concentrations of 0–2 mg/L, such concentrations are too low to be effective.

Use of Membrane Technologies

Ultrafiltration (UF), reverse osmosis (RO) and microfiltration (MF) are being used increasingly in the dairy industry to concentrate/fractionate milk for the manufacture of high moisture cheeses in high yields. Ultrafiltered retentates are commercially used to produce pre-cheese for the manufacture of some soft and semisoft cheeses, and to prepare a cheese base for processing. However, the use of highly concentrated milks (>5 times the concentration factor) for the manufacture of semi-hard and hard cheeses remains a challenge as sensory attributes of the finished cheese differ from that made from regular milk by traditional methods. Low concentration ratio ultrafiltration has been used successfully for the manufacture of Cheddar and related types. Microfiltration of milk using membranes of pore size ranging from 0.01 to 10 μm is used for the removal of bacteria and bacterial spores from milk. Membrane technologies are also used to produce new liquid and dry ingredients for use in standardizing cheese milk.

Standardization of Milk

Milk is standardized by adjusting the casein to fat (C/F) ratio in the milk to control the percentage of fat in the dry matter, produce consistent cheese and meet the regulatory standards of a given variety. Methods for increasing the casein content of cheese milk include the addition of skim-milk powder, condensed skim, milk protein concentrate (liquid or powder) or caseinates or the removal of fat as cream. Conversely, addition of cream increases the fat content of cheese milk. The most desirable C/F ratio in milk used for the manufacture of Cheddar and Gouda-type cheese is 0.7 although ratios as low as 0.64 may be desirable in some cases. Under ideal cheese making conditions, typical fat recovery would be around 93%, although under practical conditions, fat recovery is less than 93% and independent of the C/F ratio in milk. Table 6 shows that standardization of whole milk with milk protein concentrates increases cheese yields and dry matter recovery in reduced-fat Cheddar cheese.

Table: Effect of standardization of whole milk with skim milk or milk protein concentrate (MPC) on the fat recovery, dry matter recovery and yield of reduced-fat Cheddar cheese.

Treatment	Actual yield (%)	45% moisture adjusted yield (%)	Fat recovery (%)	Dry matter recovery (%)
Whole milk + skim milk	7.40 ± 0.56	8.05 ± 0.52	91.85 ± 2.68	42.25 ± 2.82
Whole milk + MPC (1% starter)	18.62 ± 2.89	19.96 ± 1.89	93.77 ± 1.83	61.21 ± 4.66

Whole milk + MPC (2% starter)	18.72 ± 2.47	20.15 ± 2.11	93.42 ± 3.18	61.70 ± 4.15
P-value	0.00	0.00	0.64	0.00

Cheese Yield

During the conversion of milk into cheese curd, milk constituents are separated into two groups, those that are retained in the curd and those that are lost in the whey. Cheese curd retains most of the fat and casein in the original milk while the whey contains mostly water, lactose, proteins (peptides and other nitrogenous compounds) and minerals that are soluble at the pH of cheese making. The typical yield of cheese ranges from 9 to 15% depending on the chemical composition of the milk, efficient recovery of fat and casein in the cheese, losses of milk constituent in the whey resulting from milk handling and treatment and cheese making procedures, and the final moisture content of the cheese.

The chemical composition of the milk is obviously influenced by factors such as climate and season, animal feed, age and breed of cow, stage of lactation and disease state of the animal. The composition of cheese milk resulting from membrane concentration and fractionation of milk (e.g. ultrafiltration, microfiltration and reverse osmosis) also influences cheese yields. Hence, the economics of cheese making is to maximize yields through the efficient recovery of milk constituents while minimizing constituent losses in the whey. Large scale manufacturing plants can also increase vat throughput and maximize yields by fortifying milk with milk solids to 14–17% dry matter content. In addition to yield increases, fortification reduces plant labour costs by increasing the amount of cheese per vat per man-hour. Although the use of high solids milk offers attractive returns to manufacturers, equipment to handle the volume of milk and weight of cheese presents challenges to the industry.

Pasteurization of Milk

Cheese milk must be adequately pasteurized (72 °C × 15 s). In several countries, the use of raw milk for cheese making is still prevalent. In the USA, cheese manufactured from raw milk must be stored at a minimum of 1.7 °C for at least 60 days before consumption. This regulation limits the manufacture and sale of unripened soft cheeses from raw milk. Fox's group recently carried out several studies to understand differences between raw and pasteurized milk cheeses. Cheeses made from raw milk contained higher numbers of nonstarter bacteria compared to cheeses made from pasteurized or microfiltered milk.

Optional Ingredients

Following standardization and pasteurization, optional ingredients such as calcium chloride and cheese colour (annatto) may be added to cheese milk. Calcium chloride plays a role in the second stage of milk coagulation, i.e. gelation. The legal limit in the USA for C_aCl_2 is 0.02% (w/w) milk. Annatto colour extracted from the plant Bixa Orellana is added to give uniformity in colour of certain varieties, such as Cheddar.

Starter

Various starter types are used for cheese making. Starters play a role in the acidification of cheese milk to the desired pH during manufacture. In addition, starter bacteria play an important role in the maturation and flavour development of cheese. Current starter technologies include genetically modified starters, adjunct starters and fast-acid starters, which are available commercially as liquid, frozen or dried. In large commercial factories, liquid starters are propagated daily from bulk starter vessels using internal or external pH-controlled starter media before inoculating into milk. The starter medium may also be whey-based.

Smaller, less sophisticated plants propagate starters in sterile milk or reconstituted skim milk powder. Frozen cultures are available in concentrated or un-concentrated forms and inoculated directly into milk (also called direct vat cultures). Frozen cultures may also be available as frozen pellets that are easier to dispense. Dried starters may be spray-dried or freeze-dried. Immobilized cell technology is also possible but it has not been applied commercially. Fox's research in the area of starter culture technology focused on the role of starters in cheese flavour development and the contribution of indigenous microflora and nonstarter bacteria to flavour development.

Coagulant

Chymosin is the principal milk-clotting enzyme used in cheese making. There are several coagulants of animal and microbial origin. These include calf chymosin, bovine and porcine pepsins, Mucor meihei protease, Mucor pusillus protease, Cryophonectria prasitica protease and fermentation derived chymosins. These enzymes, often called rennets, hydrolyse the micelle-stabilizing κ-casein at or near the Phe_{105}–Met_{106} bond. The rennetaltered milk coagulates in the presence of Ca^{2+} at approximately 30 °C. Rennet-hydrolysed κ-casein gives a hydrophobic para-κ-casein (f1–105) that remains with the curd, and a hydrophilic (glyco)macropeptide (f106–169) that is lost in the whey. The activity of chymosin and other milk clotting enzymes is pH dependent, with activity increasing as pH decreases.

Cooking, Whey Drainage and Salting

Following coagulation and cutting of curd, the curd is cooked, or scalded, to expel whey through syneresis. Cooking also facilitates starter growth and acid development. Salt is added to cheese to slow down or inhibit starter activity and to enhance the taste of cheese. The concentration of salt also affects enzymatic activity in cheese during cheese ripening. The salt content in cheese is best expressed as salt-in moisture (S/M), which controls the extent to which starter activity continues after salting. Typical S/M levels in Cheddar-type cheeses are 4–6%; beyond 6% S/M the activity of most starter bacteria is inhibited.

Cheese Ripening

Cheese ripening is an outcome of several biochemical and metabolic processes usually referred to as glycolysis, lipolysis and proteolysis. The relative importance of each of the processes depends on the variety of cheese. Fox has made an outstanding contribution to the understanding and advancement of cheese ripening. He has researched the role of coagulant, indigenous milk enzymes (e.g. plasmin) starter and nonstarter bacteria in proteolysis during cheese ripening.

Proteolysis is the most complex process, the extent of which depends on the variety—from very limited in mozzarella to very extensive in blue mould varieties. The general reaction steps are:

- Initial hydrolysis of caseins by residual coagulant and plasmin to large peptides.
- Breakdown of large peptides by starter proteinases and peptidases into medium and small peptides.
- Further hydrolysis of the medium and small peptides by starter peptidases into dipeptides, tripeptides and free amino acids.

While the level of residual coagulant in cheese is dependent on the pH of milk at setting the plasmin activity is dependent on the cooking temperature used during manufacture, being higher in high-cook cheeses (e.g. Swiss) than in low-cook cheeses. The peptides produced by the action of residual coagulant and plasmin are often either tasteless or bitter and do not contribute directly to the typical taste of cheese. However, the mixture of small peptides and amino acids directly influences the taste and mouth feel of cheese. The free amino acids may also be further catabolized into flavour compounds that are unique for each cheese and are dependent on the types of enzymes and microorganisms (particularly nonstarter lactic acid bacteria, NSLAB) present in cheese. The contribution and importance of NSLAB to cheese ripening has been the subject of intense research.

Flavour compounds in cheese are concentrated in the water-soluble fraction. Hence, methods for studying proteolysis and the fractionation and characterizing water-soluble nitrogen (WSN) are important in cheese research and technology. Methods for electrophoretic examination of cheese, extraction, fractionation and quantification of WSN and measurement of free amino acid content by the cadmium–ninhydrin reagent are routinely used for the studying cheese proteolysis.

References

- Food-processing, biology: byjus.com, Retrieved 20, June 2020
- Food-technology: aksuniversity.ac.in, Retrieved 04, March 2020
- Definition-food-manufacturing-6623871: bizfluent.com, Retrieved 18, April 2020
- Light-induced-reactions, food-preservation-50551: britannica.com, Retrieved 08, August 2020
- Commonly-used-food-processing-methods: discoverfoodtech.com, Retrieved 25, Jan 2020
- Pickled-fruits-and-vegetables, food-preservation-50575: britannica.com, Retrieved 10, Feb 2020

4

Processed Foods

Technology used in making processed or packaged foods is known as the technology of processed foods. Such foods include cakes and biscuits. Cakes can be processed using the blending method and boiled method. This chapter discusses in detail the technology associated with the production of these types of processed foods.

Technology of Biscuits Manufacture

The term 'Biscuit' refers to the small round bread leavened with baking powder or soda and covers a large variety of sweet, salted, filled and coated biscuits. Biscuits are eaten by all sections of people across the board round the year. They are, thus, mass consumption items with number of varieties and shapes. The market is scattered. There are some dominant national and regional brands. Biscuits can be manufactured at a location which is close to the market. The basic ingredients of biscuit are flour, baking powder, salt, baking soda, sugar, shortening and milk solids.

Biscuits including wafer biscuits shall be made from maida, vanaspati or refined edible oil or table butter or desi butter or margarine or ghee or their mixture containing any one or more of the following ingredients, namely: Edible common salt, butter, milk powder, cereals and their products, cheese cocoa, coffee extract, edible desiccated coconut, dextrose, fruit and fruits products, dry fruit and nuts, egg, edible vegetable products, ginger, gluten groundnut flour, milk and milk products, honey liquid glucose, malt products, edible oilseeds, flour and meals, spices and condiments, edible starches such as potato starch and edible flours, sugar and sugar products, invert sugar, jaggery, protein concentrates, vinegar and other nutrients and vitamins.

Types of Biscuit

Classification of biscuits on the basis of chemical composition and processing methodology is difficult due to overlap. These are manufactured in varieties covering sweet, salted, filled and coated biscuits.

BIS Classification

The varietal differences of biscuits can be easily distinguished based on sensory attributes. This standard broadly classified the biscuits on the sensory basis into the following five types:

- Type I: Sweet
- Type II: Semi-sweet
- Type III: Crackers
- Type IV: Cookies
- Type V: Speciality Biscuits

Biscuit has nowadays become a food item of mass consumption and has got wide acceptance among all the levels and class of consumers.

UK Classification

Biscuits are classified based on the formulation as follows:

S.N.	Type of biscuit	Description
1.	Hard dough biscuit:	Dough used is similar to bread dough. Mixing of flour is more to maximize gluten development. Dough is elastic and extensible due to well-developed gluten network during mixing. Very low fat and sugar content relative to flour. Example: Crackers, sweet and semisweet biscuits.
2.	Short dough biscuit:	Dough more closely related to cake dough however water used is lesser than cake dough. Mixing of flour is very little to minimize gluten development. Higher fat or shortening than hard dough resulting into lower extensibility of gluten and hence easy breaking of biscuits. Higher sugar in the formulation.
3.	Soft dough biscuit:	A subgroup of short dough biscuit with a more soft consistency which can be attributed to its higher sugar and fat content.

USA Classification

This classification of biscuits is based on the method of processing and, in particular on the basis of the way biscuits are shaped.

- Sheeting or cutting type: Here the dough is passed through laminators to produce sheets. These multilayer sheets are then fused into single sheet of desired thickness. This sheet is then passed onto cutter of various shape and sizes of cups/dies. The examples are cream crackers, puff crackers, marie gold, salt n sugar cracker, fiftyfifty, krackjack, monaco etc.

- Rotary molding type: The dough is in short crumb form and is directly fed in to rotator moulder with dies of various shape and size. The examples are Good day, Glucose chocochips biscuit etc.
- Wire cutting type: Here soft short dough is passed through holes in extruders and these doughs are then wire cut in pieces which are directly dropped into oven band. Two three types of dough can be co-extruded from the die where the central dough could be jelly or jam.These ropes can be cut after or before baking. Examples are butter cookies, chocolate cookies.
- Depositing type: Here the soft pourable short dough is passed out from extruder and directly laid on baking bands. The depositor may rotate to make swirls and twists on to the products. Cream treat biscuits are made by this method.

Biscuit Manufacturing Process

There are various established processes commercially available and being practiced by different manufacturers. The process of manufacturing of biscuits mainly involves the following three steps: Mixing, Shaping or forming and Baking. As the mixing and baking steps are common to manufacture of all types of biscuits the differences in the finished product arises due to the different shaping and forming methods.

Mixing

Mixing is commonly defined as a process designed to blend different ingredients into a uniform, homogenous mixture. The major ingredients are flour, fat, sugar and others as per the desired final one would like to have. For this step, all ingredients are put together in right proportion for dough formation. These are then fed into mixers where mixing is done and dough is prepared for moulding/cutting. There are three types of mixers being used for mixing of ingredients: a) Vertical spindle mixer, b) Horizontal mixer and c) Continuous mixer.

Vertical Spindle Mixer

- Portable dough trough.
- Spindle with horizontal paddle or arms.
- Dough is mixed by upward and downward movement.
- Mixer blades for cutting action rather than kneading.
- Advantageous being suitable for any product.
- Disadvantage is the slow operating speed (max. 20 rpm).

Horizontal Mixer

- Horizontal bowl mounted on a rigid frame that encloses the drive motor.
- Mixer bowl designed to tilt so as to eject the dough from the tilted position.
- Blades of variable shapes to perform cutting, scraping or kneading action.
- Advantage: High speed, continuous dough supply to processing line, complete automation, uniform mixing.

Continuous Mixer

- A rotor or screw operates within a barrel jacketed for temperature control.
- Continuous feeding of ingredients.
- Mixing is variable from gentle blending and dispersing to vigorous or high intensity kneading by varying the arrangements of different mixing arms along the length of the barrel.
- Constant supply of dough to processing line.
- Small in size relative to horizontal type.
- Disadvantage: Suitable for single purpose or similar type of products only, high cost.

Shaping or Forming

The forming process is specific for each biscuit, cookie type. There are three processes by which shaping or forming of biscuits is achieved; Sheeting and cutting, Rotary molding and Extrusion.

- Sheeting and cutting method: This method of forming is used for the production of crackers, semisweet biscuits and selected soft doughs.

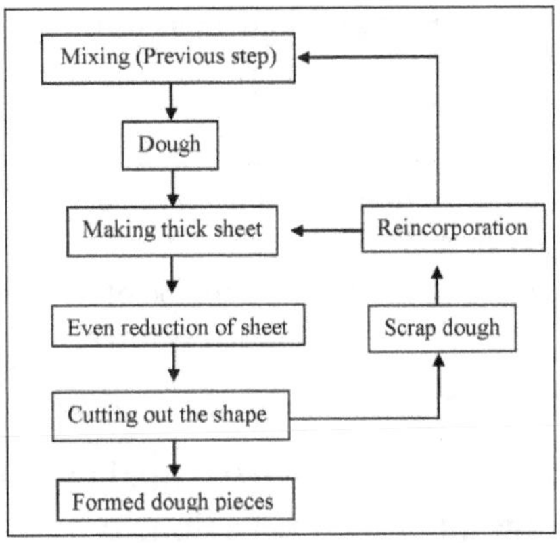

Process flow chart for Sheeting and cutting method of forming biscuits.

- Rotary molding: The rotary molding process offers advantages over sheeting and cutting that there is no scrap to recycle and very less labour requirement. This process is used for dry and crumbly doughs only.

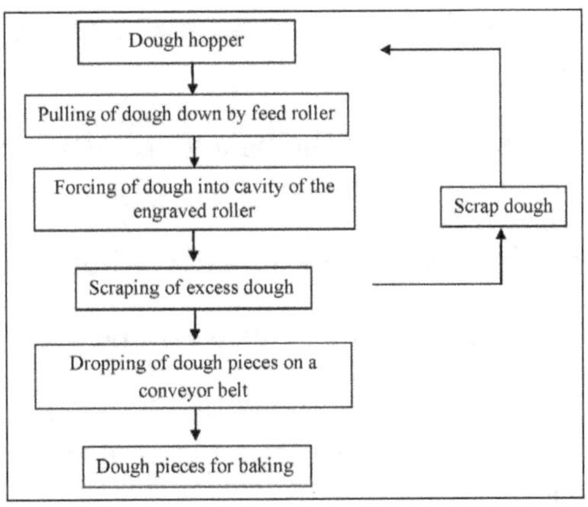

Process flow chart for Rotary molding method of forming biscuits.

- Extrusion method: In this method of dough forming, the dough is extruded through a row of dies which is cut by a wire or blade mounted on a frame that moves through the dough just below the die nozzle outlet.

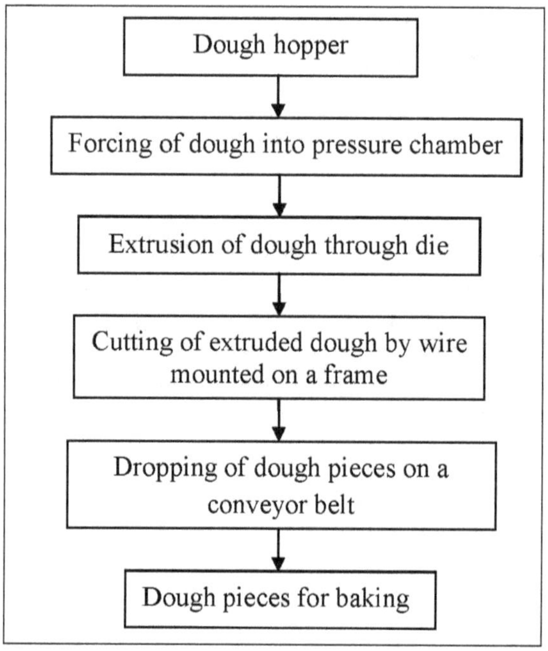

Process flow chart for Extrusion method of forming biscuits.

Baking

This is the area where we pass these moulded/formed wet biscuit into baking oven.The biscuits are baked on desired temperatures. Various type of heating are available now-a-days as per the convenience and cost. Different types of oven are available for baking products.

Cooling

The hot product obtained from baking must be cooled to room temperature prior to packaging for several reasons:

- Being warm reduced firmness of the baked product so as to withstand packaging process.
- Packaging material shrinkage due to contact to hot product.
- Hot packed product may continue release some steam causing condensation inside the packaging.
- Cooling can be achieved either by placing the baked product at ambient conditions or by forced air cooling.

Whitley Method of Biscuit Making

For the simplest biscuit preparation Whitely P.R. method is used. According to this method the following recipe is used for biscuit manufacture:

S. No.	Ingredient	Quantity (g)
1	White flour (maida)	64
2	Sugar (powdered)	18
3	Shortening	16
4	Skim milk powder	1
5	Glucose	1
6	Ammonium bicarbonate	0.5
7	Baking powder	0.3
8	Sodium bicarbonate	0.2
9	Common salt	0.4
10	Vanilla	0.02 (ml)
11	Water	as required for proper consistency (ml)

The various steps of biscuit manufacture like ingredient mixing, cutting, baking and cooling are described as follows:

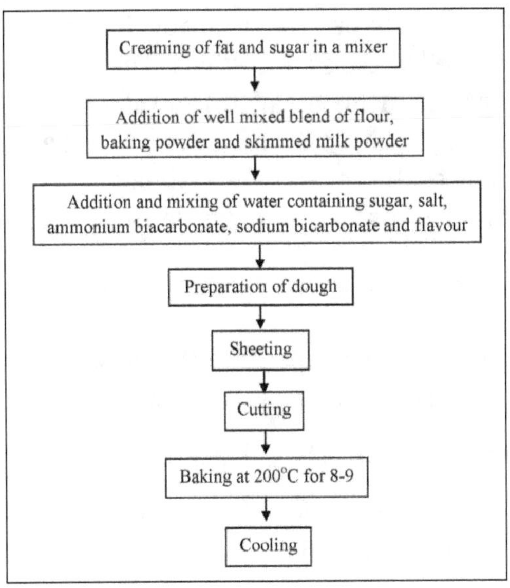

Machineries used in Biscuit Manufacturing Process

For the manufacture of biscuits the following machines are primarily used by the baking industry:

- Flour Shifter

- Stainless Steel Mixer
- GMS Churner
- Fat Churner
- Cotton Conveyor
- Laminator
- Teflon coated cutting, stamping & punching machine
- Electrically/Diesel operated baking oven
- Cooling conveyor
- Packing & sealing machine

Technology of Wafer Biscuits

Wafers are low-moisture-baked-foods. The primary textural attributes of products, the crunchiness or crispness is highly affected by the product's physical state (glassy or rubbery). Wafers represent a specialized type of light-textured biscuit, generally made from cereals. The thin, crisp, precisely shaped wafers are available in variety of shapes including flat, hollow, molded cones, rolled cones and sticks. The wafer sheet itself is baked from a simple batter containing little or no sugar. It is a rather tasteless product which has a smooth surface. Wafers usually serve as the edible carriers of an added food. However, the crispness of the wafer enhances the appeal of added food, making the wafer more than an inert carrier. Compared to other biscuits, starch gelatinization in wafer is quite high.

The Characteristics of Wafer

Wafers are baked in different fancy shapes like cones, sheets, sticks etc. The main features of wafers with respect to other bakery products are:

- Wafers are very thin biscuits: The thickness lies between 1 and 5 mm.
- Wafers often have a typical "wafer pattern" on one or both surface of the product. The surfaces are smooth and precisely formed with the dimensions, engravings, logos etc.
- Wafers are cereal based low fat products. They are made of wheat flour, sometimes with combination of other flours or starches.
- The product density of wafer ranges from 0.10 to 0.25 g/cm^3. The cross section of product reveals that wafer matrix is highly aerated and primarily composed of gelatinized starch.

- Wafers by their delicate crisp texture combine well with different types of coatings and fillings like cream, ice cream, foam etc.

Classification of Wafer

Wafers may be classified in two basic types:

- No-or low-sugar wafers: After baking, these contain from zero to a very low percentage of sucrose or other sugars. The products are of flat and hollow wafer sheets, molded cones, cups, and fancy shapes.
- Higher-sugar wafers: This type of product contains over 10% of sucrose or other sugars, which are responsible for the plasticity of the hot, freshly baked sheets. These are formed into different shapes before sugar recrystallization occurs. The products are obtained in rolled sugar cones, rolled wafer sticks or tubes, and deep-formed fancy shapes.

Wafer may be following three types:

- Plain wafers: Plain wafers may be hollow or flat or in any shape desired by the purchaser.
- Sandwiched wafers: These types of wafers shall have two or more plain wafers, sandwiched with filling in between. The filling may be cream, jam, jelly, marshmallow, caramel, figs, raisins etc. The fillings shall be not less than 50% by mass of the filled wafer.
- Coated wafers: These types of wafers include both half-coated and full-coated wafers. The coating may be of chocolate, substitute fats butter scotch etc. and the fillings shall be not less than 60% by mass of the coated wafers.

Table: Compositional requirements of wafers.

Characteristics	Wafer types	
	Plain	Sandwhich/Coated
Moisture, percent by mass, Max	4.5	6.0
Acid insoluble ash (on dry basis), percent by mass, Max	0.05	0.05
Acidity of extracted fat (as oleic acid), percent by mass, Max	1.0	1.0

Ingredients of Wafer Biscuits

The ingredients of wafer biscuit preparation can be classified in two types:

- Essential ingredients: The essential ingredients for preparation of batter of wafers are a) Flour and b) Water.

- Optional ingredients: In addition to essential ingredients following ingredients may be added to wafer batter are a) baking powder and other leavening agents, b) cereals, c) chocolate, cocoa powder, coffee extract, d) citric, malic, lactic, tartaric, acetic acid e) food colours, flavouring essences, flavour improvers and fixers f) desiccated coconut, dry fruits and nuts, g) fruit and fruit products, h) salt, dextrose, liquid glucose and sugar products (gur, jiggery etc.), i) emulsifying and stabilizing agents agents, j) preservatives and antioxidents, k) enzymes –fungal alpha amylase for dough conditioning, l) Fat and shortenings, m) malt products, n) milk and milk products, o) eggs and egg products, p) spices, condiments and their extracts , q) nutrients vitamins, minerals etc. and r) yeast.

Table: Specification for using dough improvers and conditioners.

Ingredients	Name	Quantity (g/ kg of flour)
Dough improves	Ammonium persulphate	Max 2.5 g
	Potassium bromate/ Potassium iodate	Max 50
Dough conditioners	Calcium or sodium salt of stearoyl-2-lactylate	Max 2.0
	Polysorbates	Max 2.0

Wafer Oven

The oven is the heart of the process as it both forms and bakes the wafer. Originally, wafers were made one at a time in hand-held tongs heated over an open fire. The tongs consisted of a pair of strong metal plates which, immediately batter was placed between them, were lached to resist the strong forces. This technique has been mechanized and nearly all wafer ovens still work on this principle producing sheets typically 470 × 290 mm. Larger sheets are possible but it rarely go above 470 × 350 mm. The flat wafer sheets have an overall thickness of not more than 2-5 mm. The baking plates can also carry special figures (nuts, sticks, hemispheres, fancy shapes) up to depth of around 20 mm, yielding “hollow wafer sheets”.

The plate pairs are fixed to heavy carriers which are linked together to form a chain which is circulated continuously through an insulated box (oven). The heavy carriers are required not only support the plates and to keep them rigid, but also maintain their parallel positions against the steam pressures as the baking proceeds. Wafer baking ovens frequently have 32-104 pairs of plates, continuously circulating on a chain. The length of the oven chains and the number of plate pairs has steadily increased so that presently up to 120 plate pairs may form a single plant. However ovens with 41-61 plates are most common. The plates are edged with metal strips to give a closed baking mold, except for small venting channels for steam release.

The plates are heated either directly by gas or by electric heaters and wafer ovens generally operate at temperatures between 160 °C and 190 °C. The quality of wafer sheets is judged principally by their weight, surface colour and uniformity of moisture content. Texture and crispness tend to be related to one or more of these properties, but basic differences in wafer sheet thickness will affect the internal structure.

Wafer oven maintenance programme is needed periodically to clean the plate surfaces. Deposits of charred oils and sugars build up, blackening the surfaces and causing wafer release problems. Under steady baking conditions with low sugar recipes, cleaning may be needed after about 1000 hours of running. Plate cleaning is best achieved with caustic soda solution. Modern wafer-baking plates often are surface plated, e.g. with chromium for easier release and reduction of cleaning stops.

Role of Different Ingredients in Wafers Preparation

1. Flour: Gluten strength of flour is a key to finished wafer quality, so selection of flour is also important. Gluten content should be sufficient to provide strength to the wafer without making it too weak and fragile or excessively hard and flinty. Flour milled from soft wheat for biscuit production is preferred for wafer preparation. The medium protein strength (9.5%) flour is usually used for wafer preparation. Flour with protein content between 8.1% and 10.9% gives acceptable wafer sheet. Flours with increased protein content (12.8-13.2%) lead to an unacceptable product.

Flour particle size distribution is also important parameter to produce a high quality sheet. Flour with smaller particle size gives less dense, soft and friable sheet; course flour gives unacceptable and incomplete sheets. However, it is possible theoretically to make wafer from any source of starch, provided the starch is not pre-gelatinized. In wafers there is much more complete gelatinization of the starch as compared to in other biscuits. The water absorption of flour is important as variation in this property will affect the batter consistency for any given solids content.

2. Water: Water is one of the important parameter of wafer sheet preparation. Water is added to produce a convenient consistency and makes the batter pumpable. Water helps in the distribution of ingredients. As a result of evaporation during baking, water also functions as a leavening agent. Wafer formulation should include sufficient water to disperse the ingredients and to obtain a low viscosity to flow over cover plates. Flour and water ratio should be controlled as water levels affect the texture of the sheet. The quantity of water added to wafer batter is roughly 150% of flour weight. Less water decreases the completeness of the sheet; therefore, thick, heavy and unbaked sheets occur. No relationship, still now have been established between wafer quality and hardness of water. The temperature of water alters the viscosity of batter affecting the quality of the sheet. Water temperature should be around 20 °C to prevent gluten strand formation. In warm condition, if flour has a high α-amylase activity, viscosity of the batter quickly drops. The unfavourable effect on viscosity could be minimized by using batter within 10-30 min.

3. Fat and lecithin: Fat and lecithin are used to help release wafer sheets from plates. There is little or no shortening effect by fats, but microphotographs have revealed that wafer surface is smoother when fats are present in the recipe. Without fats there would be a strong tendency for wafers to stick on the plates. Certain fats are preferred than other for incorporation at the time of mixing. For convenience, liquid vegetable oils are favoured (e.g. groundnut oil, cottonseed oil or sunflower oil). The precaution should be taken to ensure that setting of fat does not occur in the mixer prior to dispersion. Lecithin is a useful addition, it is better to include a fluidized soya lecithin with the fat or oil rather than use powdered lecithin to save cost. Though, the fat content of wafers is low, the inclusion of antioxidant in the oil is necessary to retard rancidity as the surface area of wafer and the internal texture is extremely large.

4. Egg: Eggs can serve as a source of both fat and emulsifier (lecithin).

5. Sugar and milk powder: They may be added in small quantities to improve quality. The problem associated with these ingredients that both these ingredients promote wafer colouring and sticking to the plates during baking. It is believed, that staling is retarded by incorporation of sugar and crispiness is maintained for longer as the moisture content rises. Sugar, when used in ice cream cones, it is necessary to employ specially designed equipments.

6. Salt: Salt is added as a flavour enhancer and the level is usually around 0.25 units per 100 units of flour.

7. Aeration: This is most important in wafer preparation. Although bubbles of air are included during batter mixing, most of these float out of the batter before it is deposited onto the plates. If insufficient time is allowed for the bubbles to leave the batter, the density of the batter changes during the course of use ultimately affecting baked sheet weights. Leavening agents help in development of porous texture of wafers by increasing the number of gas cells in batter. Sodium bicarbonate and ammonium bicarbonate or a mixture of two can work to create chemical aeration. Ammonium bicarbonate is particularly effective. Combination of batter consistency and ammonium bicarbonate level is the best way to control batter spread and wafer weight.

Sodium bicarbonate and ammonium-bi-carbonate also affects the final pH and influence the colour development during baking. High pH of batter retards the gluten strands formation which is undesirable. Yeast is added as a means of aeration. It is most unlikely that any flavour or textural benefits are contributed to the wafer sheet; yeast cells probably form the nuclei for water vapour production which is important for the formation of a good wafer texture. Batter standing times and suitable temperatures to allow multiplication of yeast are not usually very practical, so yeast is now rarely used in batter preparation.

8. Other ingredients: The information available about flavouring ingredients of wafer is scanty. Most synthetic flavours and essential oils are prone to steam distillation and

hence are lost in wafer production. Protein hydrolysates are fairy heat stable, but they are prone to colour development. Still they may be used for preparation of savoury wafers.

Role of Different Processing Parameter in Wafers Weight and Thickness

The processing parameters that affect wafer sheet weight and thickness are discussed as below:

- Plate gap setting: The thickness of a wafer sheet is determined by the body thickness, related to the gap between the plates and also the depth of the relief or reeding in the plates. The relief cut into the plates gives not only more interest to the appearance of the wafers, but also contributes to their mechanical strength. Thus, any reeding on the top surface should be at 45° to that on the bottom. It is interesting to note that the wafer thickness is not constantly related to the mechanical gap setting between the plates. It is always thicker. The difference is between 19–30% from older slower plants and between 5–13% for newer faster plants.
- Volume of batter: The volume is adjusted to just fill the plates allowing a minimum of extrusion through the steam vents. It has been observed that excess deposit gave not only excessive waste but also increased sheet weight.
- Batter viscosity: Changes in batter viscosity affect more than one factor. Thick batters have high solids contents and do not flow well on the plates after deposition therefore a high volume has to be applied to achieve a full wafer sheet. This gives a heavier, denser and harder wafer. The flow of thick batter on a plate can be increased by adding extra ammonium bicarbonate; the gas produced blows the batter to the edges of the plate.
- Plate closure speed: Modern plants, operate with very fast latch closure mechanisms for the plates, whereas, the older ovens operated with delayed latching operation. The change in closure speed led to different wafer quality from the same batter. Faster closure gives lower weights and thinner wafers. On some ovens the final locking position of the plates and therefore the closure time can be adjusted slightly.
- Steam venting: Large areas for steam venting allow excessive losses of batter but aid in moisture removal. Whereas, small areas control losses and batter spreading better but allow development of excessive pressures. These higher pressures cause increased wear on bearings and set up strains which have bad effects on plate settings.
- Baking speed: Baking times vary between 1.5–3 minutes with 2 minutes being about average. On large plants there is a mechanical consideration and it is generally felt that 55 plates per minute is the fastest reasonable speed.

- Most ovens, even new plants, run at approximately 60% of this speed that is only 33 plates per minute. Fast baking speeds require high plate temperatures with increased tendency to shelling (wafers with extremely fragile centre texture) and appreciable moisture gradients across the sheet when they are released. This is because there is not sufficient time for the moisture to move to the steam vents around the edges. The release of the wafers from the plates is affected because some shrinkage occurs after the wafer structure sets and is the result of drying. If the wafer is not sufficiently dried out at the time of plate opening, the shrinkage may not be sufficient to allow the sheet easily to drop away. Conversely, if the drying is too much and some surface burning has occurred sticking may be experienced for this reason also. If the heat disposition across a plate is uneven some cracks may occur in the sheet immediately the plate opens and before the sheet drops away.

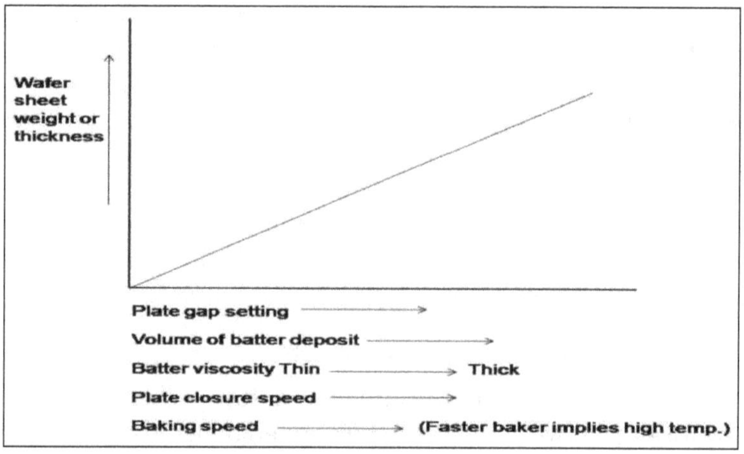

Role of different processing parameter in wafers weight and thickness.

Manufacture Technology of Wafer Biscuits

The ingredients required for preparation of Low-sugar and Higher-sugar wafer batter has listed in Table.

Table: Wafer batter ingredients (weight parts, flour = 100).

Ingredients	Low-sugar wafer	Higher-sugar wafer
Wheat flour	100	100
Water	120-160	100-140
Starch	0-12	0-5
Sucrose	0-4	25-75
Oil/ fat	0.5-2.0	1-6
Milk powder	0-2	0-2
Soy lecithin	0.2-1.0	0.2-1.5
Salt	0-0.6	0-0.6
Sodium bicarbonate	0.1-0.5	0-0.3

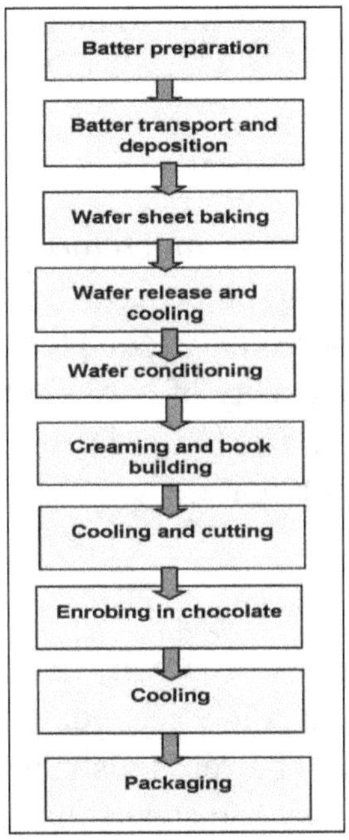

Flow diagram of wafer biscuit preparation.

Batter Preparation

Wafer batters are low viscosity batter prepared from mixing all ingredients. For preparation of wafer batter first, the water-soluble components are dissolved followed by addition of farinaceous ingredients. The batter should be of homogenous nature containing about 35-40% dry matter. Appropriate mixing is necessary for to achieve a homogenous suspension.

Mixing of Batter

The aim of the mixing process is to achieve a homogenization of ingredients and hydration of flour. Commercial mixing times range from 2.5 to 6 min. It has been observed that 4 min time is minimum required to achieve homogeneity with a good shear mixer. A high shear mixer is the best as slower mixers may allow gluten stand formation resulting the strings and lumps in the batter. The mixing process should be started as soon as possible after the assembly of all the ingredients. This reduces the possibility of a dough formation. The use of cold water also reduces the tendency for string formation.

Viscosity of Batter

The viscosity of batter should be low enough to flow over and cover the oven plate. The high viscosity results in undesirable quality of wafer. The desirable quality wafers are made from a fluid batter with a viscosity in the range of 300-2000 mPs. It was observed that hard wheat varieties produced batter with stronger rheological properties than soft varieties and the bulk elastic moduli of batter was correlated with hydration of gluten.

Greater dominance of elastic properties due to increased dispersion time of gluten led to difficulties in wafer production.

Batter Transport and Deposition

Immediately after mixing, the batter has much incorporated air, as the air rises out to mix, the batter viscosity reduces. The batter may be slightly lumpy just after preparation due to incomplete mixing. A screen is needed to remove lumps and constant gentle agitation is needed to maintain viscosity of batter. The batter is transferred just after preparation into intermediate tank. The batter is pumped to the oven from the tank via a ring main and spread on to baking plates.

Wafer Sheet Baking

The baking of wafer sheets is carried out in pairs of cast-iron metal plates with a hinge and latch on opposite sides. Within a few seconds after batter deposition, the baking molds close and locked. At first, the batter is distributed mechanically, but after some time the mold is filled completely by the steam that evolves. A small quantity of batter is extruded as waste through venting channels. The gelatinization of starch starts immediately, the pressurization of mold by steam occurs at right moment, resulting in well-aerated starch foam.

The glass temperature of wafer matrix rises, when most of the water has been driven off, resulting formation of stable structure. The temperature of wafer increases to 160-190 °C. The Maillard reactions take place at this point, resulting typical wafer colour and flavour formation. The baking times are between 1.5 and 2.5 min, depending on the wafer thickness and baking temperature. During wafer manufacturing process, there is no substantial degradation of starch molecules compared with other bakery products and extruded cereals. Therefore wafers comprise of two unique textural properties:

- Extreme crispness on biting and initial chewing.
- Good mouth feeling during prolonged chewing and swallowing owing to the absence of sticky, glutinous stimuli.

Wafer Release and Cool Down

At the end of the baking operation the plates of the oven open to release the baked sheets. The fresh batter is spread on the plates and reclosed very quickly. The sheets are cooled to room temperature by passing over an arch-type sheet cooler.

Wafer Conditioning

Wafers are passed to a conditioning unit, where the moisture content of the sheets is carefully increased so as to achieve some stability in both the texture and size of the wafer. Conditioning is particularly important for wafers that are to be chocolate enrobed. Conditioning usually involves the deliberate addition of moisture to the wafer either by storage in a humid room or by passing the sheets through a high-humidity chamber. In some cases water is actually sprayed onto wafer.

After baking, the residual moisture is 0.8-1.5%. As both the water activity of the air in the production area and the water activity of the air in the production area and the water activity of filling creams or coatings are well above that, wafers pick up moisture very easily. In line with this sorption, the dimensions of the sheet increase by 0.2-0.3% for every 1% of additional moisture. This can result in cracking of the coating in enrobed wafer biscuits. To compensate for low water activity, humidity conditioning up to ~4.5% wafer moisture is possible. This is specially recommended if enrobed or chocolate molded wafer products are made to, to avoid cracking of chocolate during its shelf life. Moreover, with increasing water activity, the wafer texture changes from a soft to a harder crispness, accompanied by a higher mechanical stability, which is good both for handling and for the final product texture. Up to 5-6% moisture content, the wafer sheets keep their typical crisp texture, but higher moisture levels will result in most cases in inadequate, tough, or even soft and soggy textures.

Table: Effect of water activity and moisture content on wafer texture.

Wafer condition	Water activity	Moisture (%), approx.	Wafer texture
Freshly baked	<0.1	<2	Very tender, crisp
After conditioning	0.3	4.5	Crisp, harder
Limit of crispness	0.5/0.55	≥ 6	Crisp to tough
Wafers, foam-filled	0.7	≥ 12	Soft to flexible
Collapse of structure	>0.85	> 20	Very soft, shrinks

Creaming and Book Building

The wafer sheets are passed to the creaming station, where a layer of a sugar and-fat-based cream is applied to one side. The most common type of cream flavours are chocolate, vanilla, hazelnut, milk, strawberry, and lemon. Creaming is done at temperatures of 30-40 °C either by contact spreading with a cream coated roller or by depositing a performed cream film. Bonding between wafers and cream is improved if a warm soft cream is applied rather than cooler stiffer one. As the cream is much softer than biscuit sandwitch cream, handling with pumps and pipes, possibly in a ring main arrangement, is possible.

Coated sheets are built up into piles as desired and a plain topping sheet is added finally. Typically there will be 3 or 4 wafers with 2 or 3 layers of cream respectively in a pile or book. The assembled book must then be pressed together before cooling. For hollow wafers, the cream is added to the hollow parts either by spreading or by controlled single depositing. Two hollow wafer sheets or a hollow and a flat sheet can be combined to form a book.

Cooling and Cutting

The wafer books pass into a cooling tunnel to set the cream. The convected cool air can also be applied. The humidity of air should be kept as low as possible to avoid moisture pick-up by the exposed wafers. The cooled books are cut into eating-size squares, rectangles, fingers, etc. by cutting them in small piles through sets of taut wires, blades or circular saws. There are two cuts at 90° to each other and resulting piles of wafer pieces are ready for packaging, storage or chocolate coating.

Packaging

The wafer biscuits have be to protect against humidity, oxygen and light to prevent oxidative deterioration and to insure a shelf life of 6-9 months. Inadequate packaging-film moisture barriers and bad sealing are the most frequent reasons for latter complaints by customers. Laminated or specially coated films are used for wafer packaging in flow packs, boxes and bags.

Molded Wafer Cones

The molded wafer cones are hollow wafers in shape of cups and fancy shapes with up to 185 mm in length. They are mainly of two types, a) No-or low-sugar cones, generally known as cake cones b) molded sugar cones. Low-sugar cones are generally similar to those for sheets and molded sugar cones have intermediate sugar content, about below 20 parts of sucrose for 100 parts of flour. The traditionally molded a cones have flat tops and regular, symmetric reedings, now a days molds for more sophisticated products are available. The cones and cups with curvilinear tops and artfully designed outsides are available in market in recent times.

The products are produced in cast-iron molds, holds for four to six items are provided in each of these molds, and 12-72 molds are circulated in one oven. The lower part of mold is made of two symmetric halves that open to release the baked pieces. The baked cones are cooled down and stacked for packaging.

Rolled Wafer Cones

Rolled sugar cones are the product containing more than 20% concentration of sucrose or other sugars in finished product. The preparation procedure of rolled sugar cone manufacturing are rather similar to no-or low sugar wafer sheet baking with requirement of reduced steam pressure. The deposit is just to form an oval circular sheet. The preparation steps comprise of batter preparation, batter transport and deposition, wafer sheet baking and wafer take-off, rolling and cone release. After baking, the mold reopens; the sheet is automatically stripped off the plate and rolled immediately on tapered mandrels to form the finished cone. A series of rolling devises mounted on round table operates continuously: sheet removal, rolling, release etc.

Rolled Wafer Sticks

Rolled wafer sticks are hollow tubes with walls consisting of very thin multiple layers, similar in texture to crunchy cereals. These layers do not carry a wafer pattern and are ~0.5 mm thick. Rolled wafer sticks have a unique brittle wall texture without being too hard in biting and chewing. The sticks are typical higher sugar wafers. A strip of batter is applied to a heated drum and baked into a continuous wafer band. The band is rolled immediately while hot into a continuous still formable tube with an internal diameter of ~6-36 mm. From here, a large family of finished product can be created. The tube is automatically cut into wafer stick pieces and cooled. These are consumed as sweet snacks or with ice cream. These can be used as intermediate products of confectionary and other preparations. The tube can be coated inside with compound chocolate or filled partially or fully with chocolate or cream. Cutting and cool down can be followed by an optional coating or decorating process.

Innovations in Wafer-Manufacturing Equipment

In recent times some new technical concepts have seen in wafer-manufacturing equipment:

- The oven with vertically stacked baking plates. The floor area requirement is less in this machine. The weight of the stacks eliminates the need for hinges as compared to traditional oven.
- Rolled wafer stick ovens with "low-emission" heating concept together with a 60% reduction in energy consumption has come into operation. Application of induction heating and replacement of gas-heated drum by a ring is the main key of the instrument. The resultant output is more consistent and controllable product quality with huge reduction in energy consumption for baking.

Wafers of different variety.

Technology of Cake Manufacture

Baking industry is the most stable sector in the food manufacturing industries. The principal basis for most products in this range is wheat flour. Addition of water and various other ingredients give us variety of products like cookies, crackers, cakes, pastries and biscuits. Biscuits are flat, crisp, baked good, whereas cookies are softer and thicker. Cookies and cakes are the some of the oldest snack foods known to the human. These types of flour confections are palatable, easy to carry and don't require further preparation before consumption and thus they are considered as staple snacks.

Product Characteristics

Cake is made by the formation of batter from soft/weak wheat flour. Cake is a traditional centerpiece of festivities and joyous celebrations. Cakes are relatively higher both in sugar and shortenings. Cake is a complex emulsion and foam system with appreciable amount of incorporated air as a small bubble into the batter. Cakes are products leavened mainly by baking powders, sometimes by air incorporation, and occasionally by yeast. Although difficult to define, cakes can be divided into two categories:

- First, cakes that are higher in fat and whose structure depends on fat-liquid emulsion created during batter mixing (e.g. fruit slab, Dundee, ginger cake). These types of cakes are characterized by high level of sugar in the formula in which starch gelatinizes during baking. Cakes set when baked giving a light product.
- Second, those with less fat (un-shortened cake), or even none at all, but rich in eggs that can aerate to a foam during mixing and gives a characteristic spongy crumb to product such as angel food cake, Victoria sponge.

Ingredients

The Flour, sugar, shortening and eggs or skim milk powder are essential ingredients. Whereas baking powder, milk, fruits and flavouring substances are optional ingredients.

- Flour: It builds the structure of cake and holds other ingredients together in an evenly distributed condition in the cake. Flour for cake making should have protein content of 7 to 9 per cent. Short patent flour with fine granule structure made from soft wheat is ideally suited for cake making. Cakes flours are bleached to a greater degree in order to brighten the colour. Bleaching also modifies the gluten forming properties (starch gelatinizes faster).
- Sugar: Sucrose is most widely used sweetening agent in cake making. Finely granulated sugar ensures even grain and soft texture in cakes. Sugar has tenderizing action on flour proteins and makes the cake tender. It helps to retain moisture in cakes and improve its shelf-life. The golden crust colour of cake is due to caramelization of sugar.

- Shortenings for cakes should have good creaming and emulsifying properties. Fats have a tenderizing action on flour proteins and thus make the cake tender by holding air cells incorporated during creaming process. It also acts as a moisture retainer and helps to keep the cake moist and thus improves the shelf-life of cakes. Fat used in cake making should be of plastic nature which could incorporate and hold minute air cells during creaming operation.
- Eggs and flour form the necessary framework to support the cake structure. Eggs provide moisture to the cakes. Lecithin of egg yolk acts as emulsifier and later adds to colour. Egg improves taste, flavour and nutritional value.
- Milk adds richness and structure to the cake: Milk proteins have binding action on flour protein which creates toughness in cakes. Milk sugar lactose improves the crust colour, moisture retention property and flavour. Milk solids also improve the nutritive value.
- Water: Formation of gluten, release of CO_2 gas from baking powder and formation of vapour pressure are made possible by presence of water. Water regulates the consistency of batter which affects the volume and texture of the cakes.
- Salt enhances the natural flavour of other ingredients used in cake making. It also controls development of crust colour by lowering the caramelization temperature of sugar. It may be used at the rate of 0.7 to 1.2% depending upon flavour.
- Baking powder of various types, when moistened with water and heated, evolve CO_2 gas which expands during baking and impart volume to cakes.
- Flavourings: Cocoa, chocolate, vanilla etc. are added as flavours.

There are different methods manufacturing cake: Sugar batter method, Flour batter method, Blending method, Boiled method, Sugar water method and all in process.

Sugar Batter Method

- Creaming: In case of using different shortenings, these should be first creamed together in order to blend them thoroughly. Sugar is then added gradually avoiding the adverse effect in the aeration process. Very light texture and brighter appearance of the mixture indicates that the adequate aeration has been achieved.
- Addition of dry ingredients: The well mixed blend of flour, baking powder and other dry ingredients is added to the mixture in portions. To avoid toughening of gluten, after addition of dry ingredients there should be minimum possible mixing action. Hence mixing of batter at low speed is essential for retaining the entrapped air as well as toughening of gluten.

- Addition of liquid ingredients: When mixing of dry ingredients complete the liquid ingredients are added. Addition of liquids enhances the fluidity of batter to an optimum level required for gradual and uniform leavening of cake during baking operation.
- Baking: The batter is poured on baking pan and baked at °C for minutes. The baked cake is then cooled and packaged/consumed.

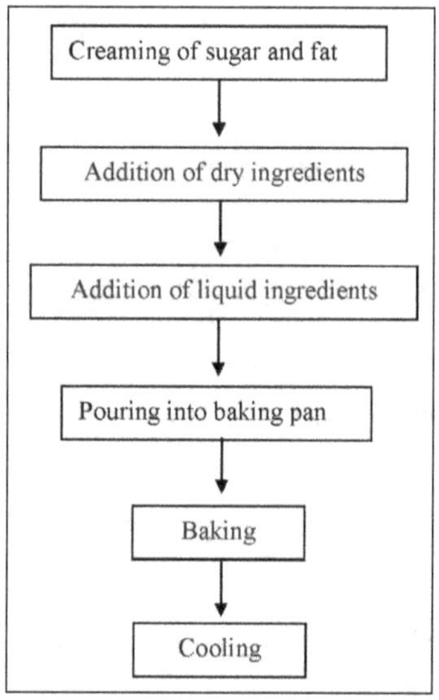

Process flow chart for sugar batter method.

Flour-Batter Method

The flour batter method is especially suited for making lean cakes which do not contain much fat or egg and most of aeration is achieved through baking powder.

- Creaming of fat and flour: Fat and equal quantity of flour is creamed together till a light and fluffy mixture. During creaming the flour is added gradually. The advantage of this step is that as a major portion of the flour is coated with fat before addition of any liquid, the gluten development is avoided when the flour is mixed with liquid. Due to the same reason slightly strong flour may also find use in cake manufacture.
- Addition of whipped mixture of egg and sugar: Egg and an equal quantity of sugar are whipped to a stiff froth. This is added in small portions to the creamed mixture of fat and flour to avoid the curdling of batter and after each addition, mixed thoroughly.
- Addition of liquids: The remaining sugar is dissolved in milk or water and added to the mixture. Any colour or flavour is also added along with this liquid.

- Addition of remaining flour: The properly mixed blend of remaining flour and baking powder is added and mixed gently. At this stage vigorous mixing may induce knocking out of air cells resulting into poor cake volume.

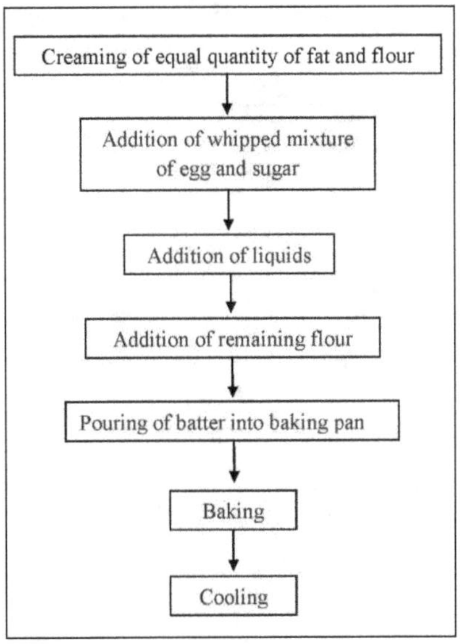

Process flow chart for blending method.

Blending Method

This is used for formulations containing more sugar than quantity of flour. All ingredients except sugar and milk are mixed together. Sugar, milk, colour and flavours are mixed and added to previous mixture followed by eggs and mixed to form a smooth batter. The blending method is used for manufacture of high sugar to flour ratio cakes.

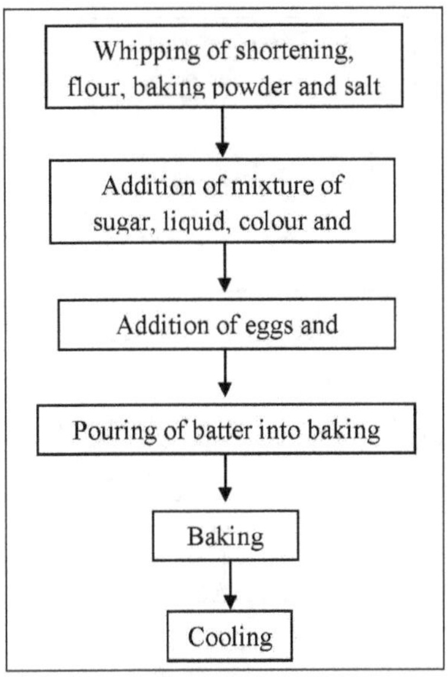

Process flow chart for blending method.

Boiled Method

Flour (more than two third portions) is added to melted butter or margarine and mixed well. Eggs are whisked with sugar followed by addition of colour and flavor. This is added to flour-fat mixture in equal parts, mixing thoroughly at each stage. Remaining flour is also added at this stage.

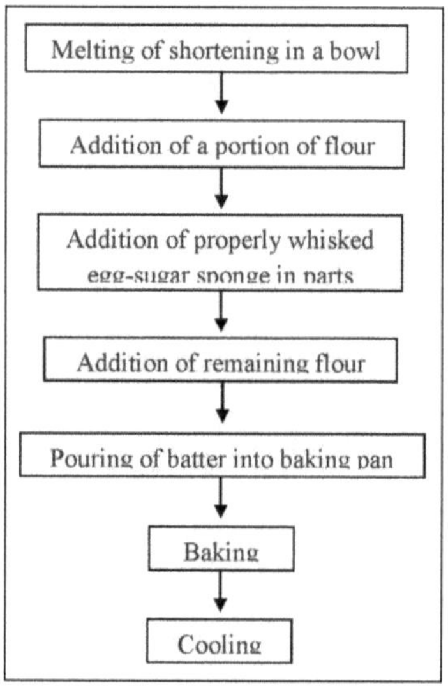

Process flow chart for boiled method.

Sugar-Water Method

The cake from sugar-water has more aeration and better emulsification resulting into better texture and longer shelf life.

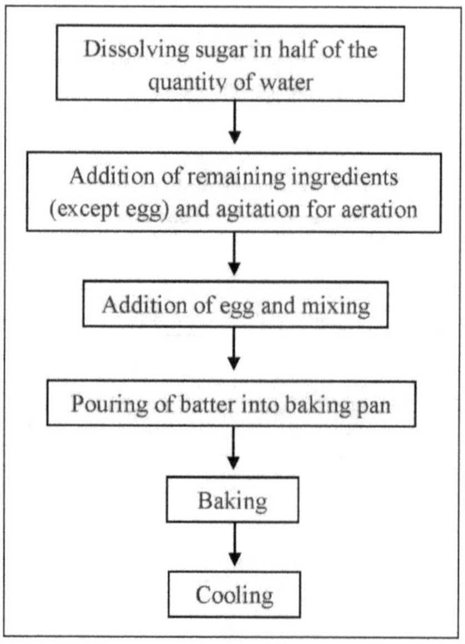

Process flow chart for Sugar water method.

All in Process

After adding all the ingredients, the mixing is controlled as follows:

- Half a minute at low speed: This is done so that all the dry ingredients are moistened without flying off from the bowl.
- Two minute at fast speed: All the ingredients break and are incorporated evenly throughout the mass. The batter is also well aerated.
- Two minute at medium speed: Aeration achieved during the second stage is not evenly disturbed in the batter. By mixing at medium speed the large air cells break up into smaller cells and the aeration of the mixture becomes even.
- One minute at low speed: This is done in order to eliminate any possible large air pockets and still finer breaking down of air cells.

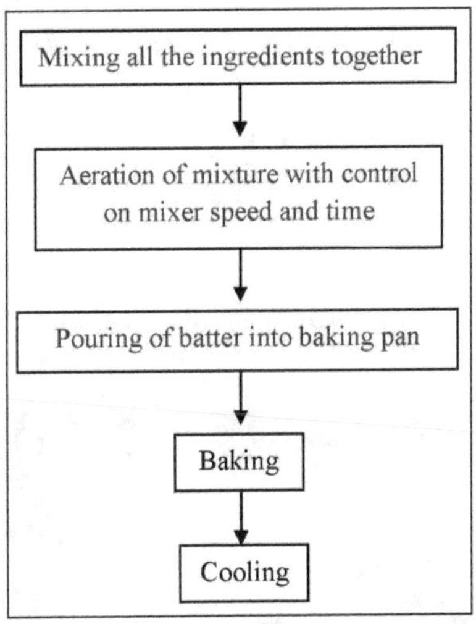

Process flow chart for All in process method.

5

Meat Products

The technology that is involved in the processing of meat products is termed as meat processing technology. It includes meat tenderization, cooking and canning of meat. The topics elaborated in this chapter will help in gaining a better perspective about the technology of meat products as well as the effects of processing and preservation technologies on meat.

Meat processing is the preparation of meat for human consumption. Meat is the common term used to describe the edible portion of animal tissues and any processed or manufactured products prepared from these tissues. Meats are often classified by the type of animal from which they are taken. Red meat refers to the meat taken from mammals, white meat refers to the meat taken from fowl, seafood refers to the meat taken from fish and shellfish, and game refers to meat taken from animals that are not commonly domesticated. In addition, most commonly consumed meats are specifically identified by the live animal from which they come. Beef refers to the meat from cattle, veal from calves, pork from hogs, lamb from young sheep, and mutton from sheep older than two years.

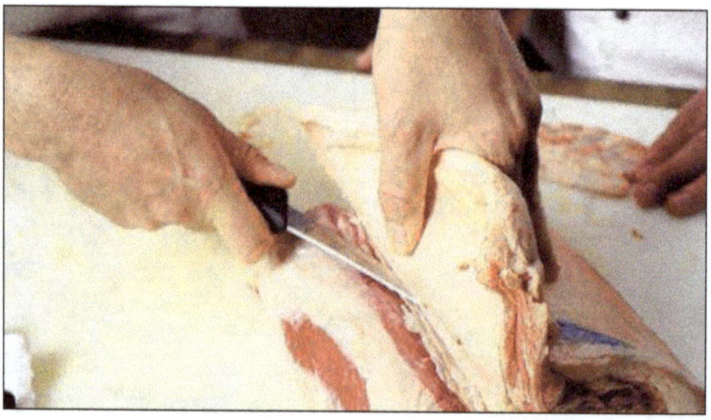

Meat cutting: A butcher cutting beef.

Conversion of Muscle to Meat

Muscle is the predominant component of most meat and meat products. Additional components include the connective tissue, fat (adipose tissue), nerves, and blood vessels that surround and are embedded within the muscles. The structural and biochemical properties of muscle are therefore critical factors that influence both the way animals are handled before, during, and after the slaughtering process and the quality of meat produced by the process.

Muscle Structure and Function

There are three distinct types of muscle in animals: smooth, cardiac, and skeletal. Smooth muscles, found in the organ systems including the digestive and reproductive tracts, are often used as casings for sausages. Cardiac muscles are located in the heart and are also often consumed as meat products. However, most meat and meat products are derived from skeletal muscles, which are usually attached to bones and, in the living animal, facilitate movement and support the weight of the body.

Skeletal Muscle Structure

Skeletal muscles are divided from one another by a covering of connective tissue called the epimysium. Individual muscles are divided into separate sections (called muscle bundles) by another connective tissue sheath known as the perimysium. Clusters of fat cells, small blood vessels (capillaries), and nerve branches are found in the region between muscle bundles. Muscle bundles are further divided into smaller cylindrical muscle fibres (cells) of varying lengths that are individually wrapped with a thin connective tissue sheath called the endomysium. Each of the connective tissue sheaths found throughout skeletal muscle is composed of collagen, a structural protein that provides strength and support to the muscles.

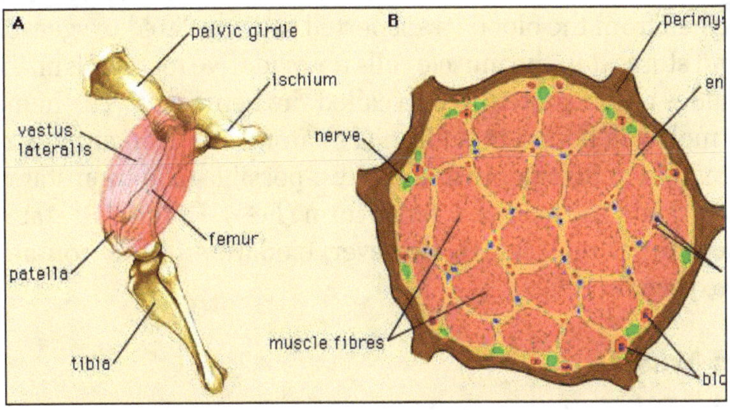

Vastus lateralis muscle of sheep.

The plasma membrane of a muscle cell, called the sarcolemma, separates the sarcoplasm (muscle cell cytoplasm) from the extracellular surroundings. Within the sarcoplasm of each individual muscle fibre are approximately 1,000 to 2,000 myofibrils. Composed of the contractile proteins actin and myosin, the myofibrils represent the smallest units of contraction in living muscle.

Skeletal Muscle Contraction

The contraction of skeletal muscles is an energy-requiring process. In order to perform the mechanical work of contraction, actin and myosin utilize the chemical energy of the molecule adenosine triphosphate (ATP). ATP is synthesized in muscle cells from the storage polysaccharide glycogen, a complex carbohydrate composed of hundreds of covalently linked molecules of glucose (a monosaccharide or simple carbohydrate). In a working muscle, glucose is released from the glycogen reserves and enters a metabolic pathway called glycolysis, a process in which glucose is broken down and the energy contained in its chemical bonds is harnessed for the synthesis of ATP. The net production of ATP depends on the level of oxygen reaching the muscle. In the absence of oxygen (anaerobic conditions), the products of glycolysis are converted to lactic acid, and relatively little ATP is produced. In the presence of oxygen (aerobic conditions), the products of glycolysis enter a second pathway, the citric acid cycle, and a large amount of ATP is synthesized by a process called oxidative phosphorylation.

In addition to carbohydrates, fats supply a significant amount of energy for working muscles. Fats are stored in the body as triglycerides (also called triacylglycerols). A triglyceride is composed of three fatty acid molecules (nonpolar hydrocarbon chains with a polar carboxyl group at one end) bound to a single glycerol molecule. If the fat deposits are required for energy production, fatty acids are released from the triglyceride molecules in a process called fatty acid mobilization. The fatty acids are broken down into smaller molecules that can enter the citric acid cycle for the synthesis of ATP by oxidative phosphorylation. Therefore, the utilization of fats for energy requires the presence of oxygen.

An important protein of muscle cells is the oxygen-binding protein myoglobin. Myoglobin takes up oxygen from the blood (transported by the related oxygen-binding protein hemoglobin) and stores it in the muscle cells for oxidative metabolism. The structure of myoglobin includes a non-protein group called the heme ring. The heme ring consists of a porphyrin molecule bound to an iron (Fe) atom. The iron atom is responsible for the binding of oxygen to myoglobin and has two possible oxidation states: the reduced, ferrous form (Fe^{2+}) and the oxidized, ferric form (Fe^{3+}). In the Fe^{2+} state iron is able to bind oxygen (and other molecules). However, oxidation of the iron atom to the Fe^{3+} state prevents oxygen binding.

Postmortem Muscle

Once the life of an animal ends, the life-sustaining processes slowly cease, causing significant changes in the postmortem (after death) muscle. These changes represent the conversion of muscle to meat.

pH Changes

Normally, after death, muscle becomes more acidic (pH decreases). When an animal is bled after slaughter (a process known as exsanguination), oxygen is no longer available to the muscle cells, and anaerobic glycolysis becomes the only means of energy production available. As a result, glycogen stores are completely converted to lactic acid, which then begins to build up, causing the pH to drop. Typically, the pH declines from a physiological pH of approximately 7.2 in living muscle to a postmortem pH of approximately 5.5 in meat (called the ultimate pH).

Protein Changes

When the energy reserves are depleted, the myofibrillar proteins, actin and myosin, lose their extendability, and the muscles become stiff. This condition is commonly referred to as rigor mortis. The time an animal requires to enter rigor mortis is highly dependent on the species (for instance, cattle and sheep take longer than hogs), the chilling rate of the carcass from normal body temperature (the process is slower at lower temperatures), and the amount of stress the animal experiences before slaughter.

Eventually the stiffness in the muscle tissues begins to decrease owing to the enzymatic breakdown of structural proteins (i.e., collagen) that hold muscle fibres together. This phenomenon is known as resolution of rigor and can continue for weeks after slaughter in a process referred to as aging of meat. This aging effect produces meats that are more tender and palatable.

Properties of Meat

Chemistry and Nutrient Composition

Regardless of the animal, lean muscle usually consists of approximately 21 percent protein, 73 percent water, 5 percent fat, and 1 percent ash (the mineral component of muscle). These figures vary as an animal is fed and fattened. Generally, as fat increases, the percentages of protein and water decrease. The Table provides a comparison of the nutrient composition of many meat products.

Table: Nutrient composition of red meats (per 100 g).

Meat type and cut	Energy (kcal)	Water (g)	Protein (g)	Fat (g)	Cholesterol (mg)	Vitamin b12 (µg)	Thiamin (mg)	Iron (mg)	Zinc (mg)
Beef									
Chuck arm pot roast	219	58	33.02	8.70	101	3.40	0.080	3.79	8.66
Rib eye steak	225	59	28.04	11.70	80	3.32	0.100	2.57	6.99
Short ribs	295	50	30.76	18.13	93	3.46	0.065	3.36	7.80
Tenderloin	212	60	28.25	10.10	84	2.57	0.130	3.58	5.59
Top sirloin	200	61	30.37	7.80	89	2.85	0.130	3.36	6.52
Ground (extra lean)	265	54	28.58	15.80	99	2.56	0.070	2.77	6.43
Pork									
Loin roast	169	62	30.24	7.21	78	0.55	0.639	1.06	2.31
Tenderloin	164	66	8.14	4.81	79	0.55	0.940	1.47	2.63
Boston shoulder roast	232	61	24.21	14.30	85	0.93	0.669	1.56	4.23
Spareribs	397	40	29.06	30.30	121	1.08	0.382	1.85	4.60

Cured ham (extra lean)	145	68	20.93	5.53	53	0.65	0.754	1.48	2.88
Lamb									
Leg roast	191	64	28.30	7.74	89	2.64	0.110	2.12	4.94
Loin chop	202	63	26.59	9.76	87	2.16	0.100	2.44	4.06
Blade chop	209	63	24.61	11.57	87	2.74	0.090	2.07	6.48
Veal									
Loin chop	175	65	26.32	6.94	106	1.31	0.060	0.85	3.24
Rib chop	177	65	25.76	7.44	115	1.58	0.060	0.96	4.49

Protein

Meat is an excellent source of protein. These proteins carry out specific functions in living muscle tissue and in the conversion of muscle to meat. They include actin and myosin (myofibrillar proteins), glycolytic enzymes and myoglobin (sarcoplasmic proteins), and collagen (connective tissue proteins). Because the proteins found in meat provide all nine essential amino acids to the diet, meat is considered a complete source of protein.

Fat

Fats, in the form of triglycerides, accumulate in the fat cells found in and around the muscles of the animal. Fat deposits that surround the muscles are called adipose tissue, while fat that is deposited between the fibres of a muscle is called marbling. In the diet the fats found in meat act as carriers for the fat-soluble vitamins (A, D, E, and K) and supply essential fatty acids (fatty acids not supplied by the body). In addition to their role as an energy reserve, fatty acids are precursors in the synthesis of phospholipids, the main structural molecules of all biological membranes.

Fatty acids are classified as being either saturated (lacking double bonds between their carbon atoms), monounsaturated (with one double bond), or polyunsaturated (containing several double bonds). The fatty acid composition of meats is dependent on several factors. In animals with simple stomachs, called non-ruminants (e.g., pigs), diet can significantly alter the fatty acid composition of meat. If non-ruminants are fed diets high in unsaturated fats, the fat they deposit in their muscles will have elevated levels of unsaturated fatty acids. In animals with multi-chambered stomachs, called ruminants (e.g., cattle and sheep), fatty acid composition found in the lean muscle is relatively unaffected by diet because microorganisms in the stomach alter the chemical composition of the fatty acids before they leave the digestive tract.

A beneficial characteristic of saturated fatty acids is that they do not undergo oxidation when exposed to air. However, the double bonds found in unsaturated fatty acids are susceptible to oxidation, and this oxidation promotes rancidity in meat. Therefore, products higher in saturated fats can generally be stored for a longer time without developing unpleasant flavours and odours.

Vitamins and Minerals

Meat contains a number of essential vitamins and minerals. It is an excellent source of many of the B vitamins, including thiamine, choline, B_6, niacin, and folic acid. Some types of meat, especially liver, also contain vitamins A, D, E, and K. Meat is an excellent source of the minerals iron, zinc, and phosphorus. It also contains a number of essential trace minerals, including copper, molybdenum, nickel, selenium, chromium, and fluorine. The Table provides a comparison of the vitamin and mineral content of different types of meat.

Cholesterol

Cholesterol is a constituent of cell membranes and is present in all animal tissues. Leaner meats typically are lower in cholesterol. Veal, however, is an exception: it is lower in fat than mature beef but has significantly higher cholesterol levels.

Carbohydrates

Meat contains virtually no carbohydrates. This is because the principal carbohydrate found in muscle, the complex sugar glycogen, is broken down in the conversion of muscle to meat. Liver is an exception, containing up to 8 percent carbohydrates.

Water

Water is the most abundant component of meat. However, because adipose tissue contains little or no moisture, as the percentage of fat increases in a meat cut, the percentage of water declines. Therefore, lean young veal may be as much as 80 percent water, while fully fattened beef may be as little as 50 percent. Because water is lost when meats are cooked, the percentages of protein and fat in cooked meats are usually higher than in their raw counterparts.

Colour

In well-bled animals approximately 80 to 90 percent of the total meat pigment is due to the oxygen-binding protein myoglobin. Colour differences in meat are related to the myoglobin content of muscle fibres and to the chemical state of the iron atom found in the myoglobin molecule.

Myoglobin Content

A number of factors influence the myoglobin content of skeletal muscles. Muscles are a mixture of two different types of muscle fibre, fast-twitch and slow-twitch, which vary in proportions between muscles. Fast-twitch fibres have low myoglobin content and are therefore also called white fibres. They are dependent on anaerobic glycolysis for energy production. Slow-twitch fibres have a high amount of myoglobin and a greater capacity for oxidative metabolism. These fibres are often called red fibres. Therefore, dark meat colour is a result of a relatively high concentration of slow-twitch fibres in the muscle of the animal. A second factor contributing to the myoglobin content of a muscle is the age of the animal—muscles from older animals often have higher myoglobin concentrations. This accounts for the darker colour of beef relative to that of veal.

The size of an animal may also affect the myoglobin content of its muscles because of differences in basal metabolic rates (larger animals have a lower metabolism). Some smaller animals (such as rabbits) typically have a lower myoglobin concentration (0.02 percent of wet weight of muscle) and lighter coloured meat than larger animals such as horses (0.7 percent myoglobin) or deep-diving animals such as whales, which have very high concentrations of myoglobin (7 percent myoglobin) and dark, purple-coloured meat. Myoglobin concentration is also greater in intact males (animals that have not been castrated) of similar age, in muscles located closer to the bones, and in more physically active animals such as game.

Oxidation State of Iron

The oxidation state of the iron atom of myoglobin also plays a significant role in meat colour. Meat such as beef viewed immediately after cutting is purple in colour because water is bound to the reduced iron atom of the myoglobin molecule (in this state the molecule is called deoxymyoglobin). Within 30 minutes after exposure to the air, beef slowly turns to a bright cherry-red colour in a process called blooming. Blooming is the result of oxygen binding to the iron atom (in this state the myoglobin molecule is called oxymyoglobin). After several days of exposure to air, the iron atom of myoglobin becomes oxidized and loses its ability to bind oxygen (the myoglobin molecule is now called metmyoglobin). In this oxidized condition, meat turns to a brown colour. Although the presence of this colour is not harmful, it is an indication that the meat is no longer fresh.

Meat Tenderization

Tenderness of meat is influenced by the breed, age, condition of nutrition and amount of muscular exercise of the animal. It mainly depends on the amount of connective tissue present between muscle fibres and to a lesser extent, on the thickness of muscle fibres themselves. Whether it's a tender juicy steak or a fall-off-the-bone tender rib, tenderness is often equated with the quality of the meat. But the texture of meat has a lot less to do with quality, and a lot more to do with how the meat is treated both before and after hitting the grill. To better understand this, let's look at the three main methods of tenderizing meat: mechanical, thermal, and enzymatic.

Mechanical tenderization involves pounding or piercing the meat with one of those medieval looking devices. The physical action is essentially pre-chewing the meat for you. I'm generally not a big fan of this technique because you're not only breaking up the connective tissues that make meat tough; you're also breaking up the meat fibers themselves.

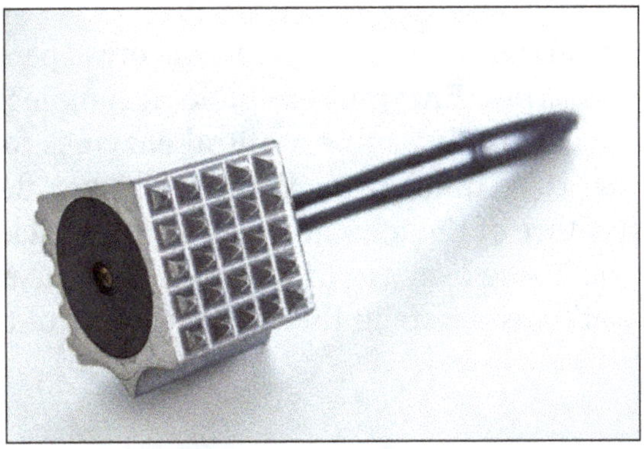

When it comes to tenderizing meat, Marc says that a meat mallet is not the only method you should consider.

While mechanical tenderization is fine if you're trying to make your meat both thin and tender, such as for Chicken Fried Steak, or Wiener Schnitzel, it's not so great for the grill. If you've ever put a tough cut of meat in a slow cooker and opened it up later in the day to find a fork-tender pot roast, then you're already familiar with thermal tenderization. In this method, heat breaks down the connective tissues, leaving you with melt-in-your-mouth tender meat.

From 140 degrees F and up, the collagen surrounding muscle fibers start shrinking and force out the juices in the meat. If you've ever had a well-done steak, you know that this makes the meat both dry and tough. But beyond 160 degrees F, the collagen begins to break down into gelatin, which not only replaces some of the lost moisture it surrounds the individual fibers with a tender gel that loosely holds the muscle fibers together. This is the stage at which meat is fork tender.

When you barbecue ribs, grill them slow and low.

The heat can be dry heat (such as from a grill), or wet heat (such as in a braise), but it needs to be applied slowly, otherwise the outer surface of the meat will start burning before the collagen in the middle has fully broken down. Thermal tenderization is great for barbecuing ribs, brisket, or any other cut of meat with a lot of collagen, but it won't do much for fillet mignon or pork chop. That's because those cuts don't have a ton of connective tissue to begin with, and you would never cook a steak or a chop long enough to breakdown the collagen they do contain.

The final method of tenderization and the one that most people are least familiar with is enzymatic tenderization. Enzymes are biological molecules that increase the rate of a reaction. In the case of dry-aging, natural enzymes found in the meat help break down the collagen over time. It results in a tender, flavorful piece of meat without altering the structure of the meat fibers. The drawback is that it takes a long time (20+ days), and you'll lose over 1/3 of the volume of meat because of moisture loss and the fact that you'll need to trim the outer surfaces of the meat before cooking it.

Preparatory Procedures of Meat

Curing

Consumers associate the majority of processed meat products like hams, bacon, and most sausages with an attractive pink or red colour after heat treatment. However experience shows that meat or meat mixes, after kitchen-style cooking or frying, turn brownish-grey or grey. In order to achieve the desired red or pink colour, meat or meat mixes are salted with common salt (sodium chloride NaCl), which contains a small quantity of the curing agent sodium nitrite ($NaNO_2$). Sodium nitrite has the ability to react with the red meat pigment to form the heat stable red curing colour. Only very small amounts of the nitrite are needed for this purpose.

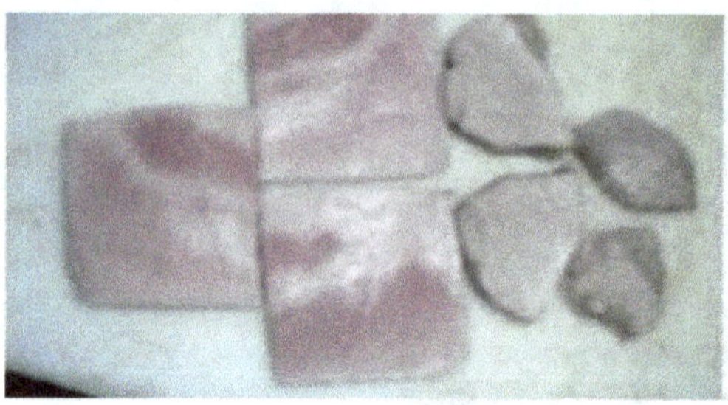	
Pieces of cooked meat (pork) 4 pieces with common salt only (right) and 3 with common salt containing small amounts of nitrite (left).	Two sausage cuts one produced with salt only (right) and the other with salt and small amounts of nitrite (left).

Nitrite can be safely used in tiny concentrations for food preservation and colouring purposes. Traces of nitrite are not poisonous. In addition to the reddening effect, they have a number of additional beneficial impacts so that the meat industries widely depend on this substance. Levels of 150 mg/kg in the meat product, which is 0.015%, are normally sufficient.

To reduce the risk of overdosing of nitrite salt, a safe approach is to make nitrite available only in a homogeneous mixture with common salt generally in the proportion 0.5% nitrite and the balance of sodium chloride (99.5%). This mixture is called nitrite curing salt. At a common dosage level of 1.5- 3% added to the meat product, the desired salty flavour is achieved and at the same time the small amount of nitrite needed for the curing reaction is also provided. Due to the sensory limits of salt addition (salt contents of 4% are normally not exceeded), the amounts of nitrite are kept low accordingly. A great deal of research has been done with regard to the utilization of nitrite and it can be said that nitrite in meat products is safe if basic rules are adhered to. Nitrite is now recognized a substance with multifunctional beneficial properties in meat processing:

- The primary purpose of nitrite is to create a heat resistant red colour in a chemical reaction with the muscle pigment, which makes cured meat products attractive for consumers.
- Nitrite has a certain inhibitory effect on the growth of bacteria. This effect is particularly pronounced in canned meat products which are usually stored without refrigeration, where small numbers of heat resistant bacteria may have survived but their growth is inhibited by the presence of nitrite.
- Nitrite has the potential of attributing a specific desirable curing flavour to cured products.
- In the presence of nitrite fats are stabilized and rancidity in meat products retarded i.e., an antioxidant effect.

Many attempts have been made to replace nitrite by other substances, which would bring about the same beneficial effects as listed above. Up to now no alternative substance has been found. As the above desirable effects are achieved with extremely low levels of nitrite, the substance can be considered safe from the health point of view. Currently the known advantages of nitrite outweigh the known risks.

Curing of Chopped/Comminuted Meat Mixtures

Curing is applied for most chopped meat mixtures or sausage mixes for which a reddish colour is desired. The curing agent nitrite is added in dry form as nitrite curing salt. The reaction of nitrite with the red meat pigment starts immediately. Due to homogenous blending the meat pigments have instant contact with the nitrite. Higher temperatures during processing, e.g. “reddening” of raw cooked type sausages at 50 °C or scalding/ cooking of other products at 70-80 °C, accelerate the process.

Another accelerating or "catalytic" effect is the addition of ascorbic acid, which slightly lowers the pH of the meat mixture. However, the dosage of ascorbic acid must be low (0.05%), just to provide the slightly acid conditions for the reduction of $NaNO_2$ to NO. A pronounced reduction of the pH would negatively affect the water binding capacity of the product which is not desirable.

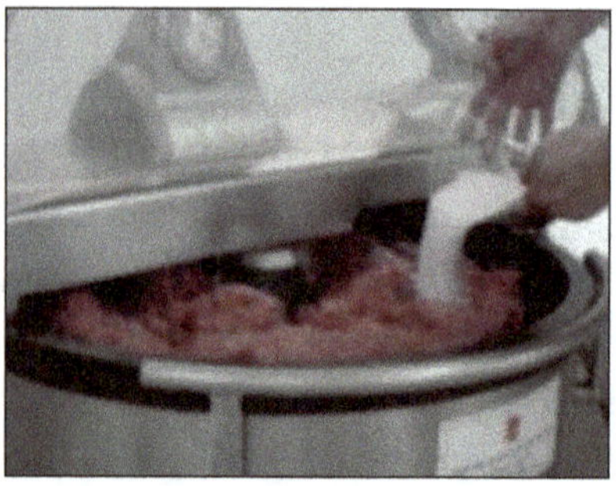

Adding of nitrite curing salt during initial phase of meat mix fabrication.

Curing of Entire Meat Pieces

Besides the curing of chopped meat mixtures, entire pieces of muscle meat can be cured. However, due to size the curing substances cannot instantly react with the meat pigments as is the case in chopped meat mixes. Hence various curing techniques are applied. The final products of curing entire meat pieces are either cured raw fermented products or cured cooked products. The curing system to be used depends on the nature of the final product (uncooked or cooked). There are two systems for curing entire meat pieces, dry curing and wet curing ("pickling") and the type of the final product determines which system will used.

In dry curing a curing mix is prepared containing salt or nitrite curing salt, together with spices and other additives. The pieces of meat are rubbed with this curing mix and packed in tanks. The curing mix gradually permeates into the meat, which can be a lengthy process ranging from several days to several weeks.

<table>
<tr><td>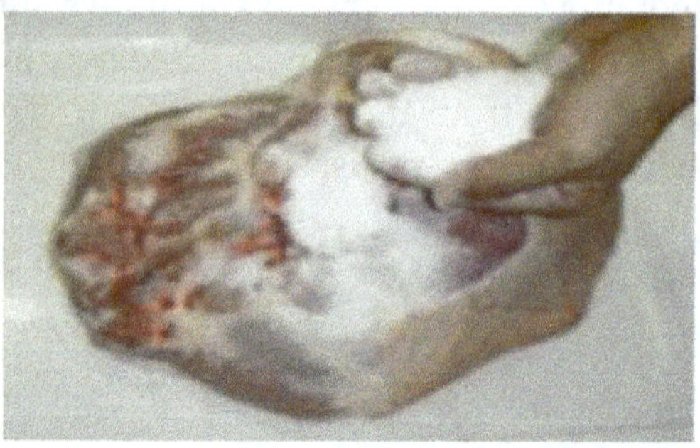</td><td>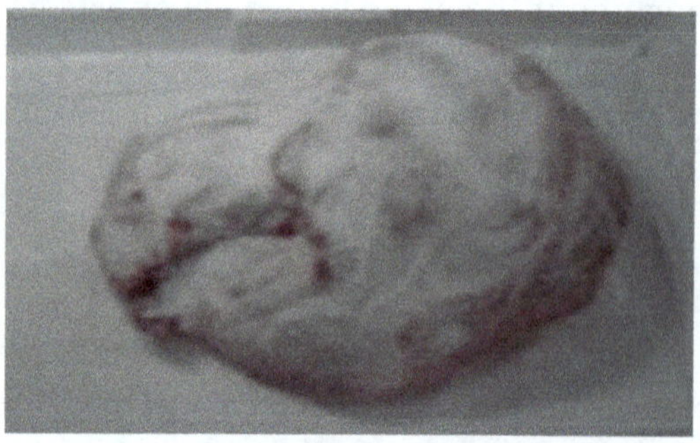</td></tr>
<tr><td>Application of dry curing mix (curing salt, curing accelerators, spices) on fresh ham (pork leg).</td><td>Ham is uniformly covered by curing mix.</td></tr>
</table>

The second method of curing meat pieces is wet curing, also called pickling, which involves the application of curing brine to the meat. For the manufacture of the brine, curing salt and spices, and other additives if required are dissolved in water. The meat cuts are packed in tanks and brine is added until all pieces are completely covered. A temperature of +8 to +10 °C for the curing room is recommended as lower temperatures may retard curing. For equal penetration of the brine, the meat is cured for periods ranging from several days to two weeks depending on the size of the cuts and curing conditions. After completion of the curing, ripening periods for the products follow for taste and flavour build-up.

Wet curing.

An alternative and quick way of wet curing is to accelerate the diffusion of the curing substances by pumping brine into the meat tissue ("injection curing"). For this purpose brine injectors with perforated hollow needles are used. The injection of brine into the muscles can be done manually by using simple pumping devices. At the industrial level semi-automatic multi-needle brine injectors are used which achieve very even distribution of the curing ingredients and can reduce the curing period (equal distribution of the curing substances or "“resting period") to less than 48 hours.

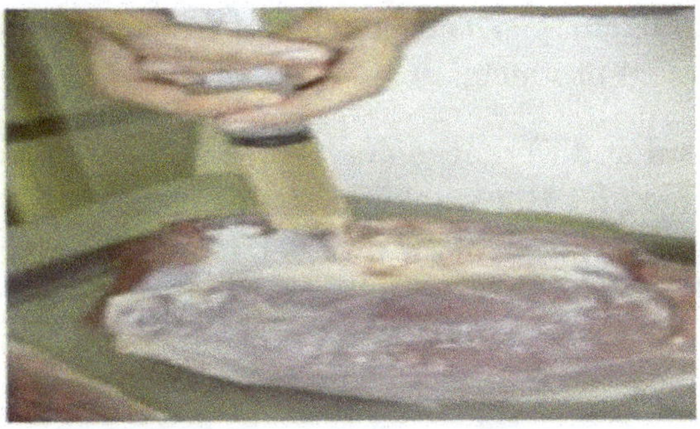	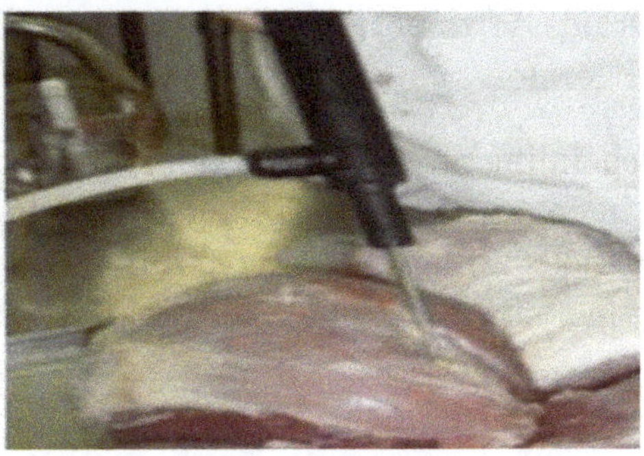
Manual brine injection using a large syringe.	Brine injection with a manual curing pump.

In addition, most injection cured meat pieces which are to be processed into cured-cooked products (such as cooked hams etc.), are submitted to a tumbling process. Tumbling further accelerates the brine penetration throughout the meat prices and "resting periods" are not necessary.

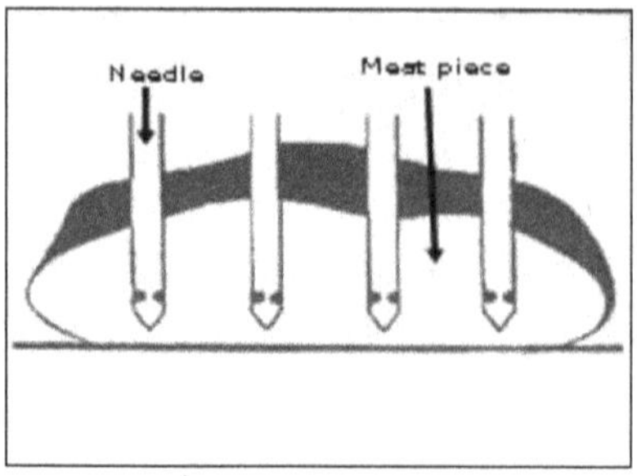

Multi-needle brine injection (principle).

Smoking

Smoke for treatment of meat products is produced from raw wood. Smoke is generated through the thermal destruction of the wood components lignin and cellulose. The thermal destruction sets free more than 1000 desirable or undesirable firm, liquid or gaseous components of wood. These useful components contribute to the development of the following desirable effects on processed meat products:

- Meat preservation through aldehydes, phenols and acids (anti-microbial effect).
- Antioxidant impact through phenols and aldehydes (retarding fat oxidation).
- Smoke flavour through phenols, carbonyls and others (smoking taste).
- Smoke colour formation through carbonyls and aldehydes (attractive colour).
- Surface hardening of sausages/casings through aldehydes (in particular for more rigid structure of the casing).

The most known undesirable effect of smoking is the risk of residues of benzopyrene in smoked products which can be carcinogenic if the intake is in high doses over long periods. With normal eating habits, a carcinogenic risk is normally not associated with moderately smoked food such as smoked meat products.

Depending on the product, smoke is applied at different temperatures. There are two principal smoking techniques: Cold smoking and Hot smoking. The principle of both methods is that the smoke infiltrates the outside layers of the product in order to develop flavour, colour and a certain preservation effect.

Cold Smoking

This is the traditional way of smoking of meat products and was primarily used for meat preservation. Nowadays it serves more for flavour and colour formation, for example in sausages made from precooked materials such as liver sausage and blood sausage.

The combination of cold smoking and drying/ripening can be applied to fermented sausages and salted or cured entire meat pieces), in particular many raw ham products. In long-term ripened and dried hams, apart from providing colour and flavour, the cold smoking has an important preservative effect as it prevents the growth of moulds on the meat surfaces. The optimal temperature in "cold" smoking is 15 to 18 °C (up to 26 °C). Sawdust should be burned slowly with light smoke only and the meat hung not too close to the source of the smoke. Cold smoking is a long process which may take several days. It is not applied continuously, but in intervals of a few hours per day.

Hot Smoking

Hot smoking is carried out at temperatures of +60 to 80 °C. The thermal destruction of the wood used for the smoking is normally not sufficient to produce these temperatures in the smoking chamber. Hence, additional heat has to be applied in the smoking chamber.

Hotdogs are placed in the smokehouse for hot smoking (pale colour before smoking).

After completion of the smoking process (brown-red colour after smoking).

The relatively high temperatures in hot smoking assure a rapid colour and flavour development. The treatment period is kept relatively short in order to avoid excessive impact of the smoke (too strong smoke colour and flavour). Hot smoking periods vary from not much longer than 10 minutes for sausages with a thin calibre such as frankfurters to up to one hour for sausages with a thick calibre such as bologna and ham sausage and products like bacon and cooked hams.

Products and Smoking

Cold smoking is used for fermented meat products (raw-cured ham, raw-fermented sausage) and precooked-cooked sausage (liver and blood sausages). Hot smoking is used for a range of raw-cooked sausages, bacon and cooked ham products. Smoke treatment can only be applied, if meat the products are filled in casings permeable to smoke. All natural casings are smoke permeable, as are cellulose or collagen basis synthetic casings.

Smoke permeable casings can also be treated using a new technology, where a liquid smoke solution is applied on the surface. This can be done by dipping in solution, showering (outside chamber) or atomization (spraying inside chamber). Polyamide or polyester based synthetic casings are not permeable to smoke. If smoke flavour is wanted for products in such casings, small quantities of suitable smoke flavour (dry or liquid) are added directly to the product mix during manufacture.

Cooking

Cooking methods in the culinary arts are divided into two categories: (1) Dry heat cooking: roasting, broiling or sautéing. (2) Moist heat cooking: braising, steaming or poaching. Because every cooking method uses either dry heat or moist heat (or sometimes both), classifying them this way ensures that every known method falls into one category or the other.

Dry Heat Cooking

Dry heat cooking refers to any cooking technique where the heat is transferred to the food item without using any moisture. Dry-heat cooking typically involves high heat, with temperatures of 300 °F or hotter. Baking or roasting in an oven is a dry heat method because it uses hot air to conduct the heat. Pansearing a steak is considered dry-heat cooking because the heat transfer takes place through the hot metal of the pan. Note that the browning of food (including the process by which meat is browned, called the Maillard reaction) can only be achieved through dry-heat cooking. Examples of dry-heat methods include:

- Roasting and Baking
- Grilling and Broiling
- Sautéing and Pan-Frying
- Deep-Frying

Moist Heat Cooking

Moist heat cooking methods include any techniques that involve cooking with moisture — whether it's steam, water, stock, wine or some other liquid. Cooking temperatures are much lower, anywhere from 140 °F to a maximum of 212 °F, because water doesn't get any hotter than that. Examples of moist-heat cooking methods include:

- Poaching, Simmering and Boiling
- Steaming
- Braising and Stewing

Roasting

Roasting works well for large cuts of beef, pork or lamb such as roasts, ribs, leg of lamb or pork loins. Depending on size you can brown the meat in a pan before roasting by seasoning with salt/pepper & oil. After browning place the meat in a roasting pan and make sure the fat side is turned up. The oven should be pre heated up to 450 °C. If you have not browned the meat beforehand than roast at 450 degree at for approximately 20 min. Then turn the temperature down to 325 degree Roasting time and temperature will vary based on the type and size of the meat entree being cooked.

Grilling

Grilling is often used for burgers, winers and sausages, steaks, ribs and loin. There are electric indoor grills as well as charcoal or gas outdoor grill. Grilling can also be used to create kabobs or to grill vegetables, including baking potatoes. You can grill over direct heat or indirect heat, depending on the size of the meat and desired tenderness. Grilling should be done carefully as to note char the meat on the outside. This can ruin the flavour of and cause the meat to be extremely tough.

Broiling

The terms grilling and broiling are often used interchangeably when it comes to using a charcoal grill. But oven broiling is different because the oven is temperature controlled. You can broil steaks, burgers and ham ect. Using a broiling pan and by turning your oven temperature to the broil settings or to a different desired temperature. While cooking, the broiler pan allows grease to drip through cracks into a bottom pan. It's a fast, healthy way to cook meat. Consider moving the oven rack to the bottom slots if possible to prevent the splashing of grease or by a special a special broiler appliance if you prefer this over using your oven.

Pan-Fry or Stir Fry

Pan frying is the method of using a frying pan to cook meat. You can add oil to some meats or simply fry the meats own fat (particularly burgers). For thicker cuts of meats, cook on medium to medium high heat for longer periods of time to ensure doneness in the mild sections of the meat. Cover the pan to trap the steam and promote through cooking. Stir frying uses a pan as well, but involves cooking small cuts of meats along with vegetables and sauces to create a delicious dish. It is often associated with oriental dishes, but can be used to create a number of creative entrees.

Stewing

In stewing, small chunks of meat in a pot or slow cooker along with vegetables and seasonings to create various soups and stews. This is accomplished by boiling and then simmering on lower heat.

- To create desired stew flavour, add stocks, water, wine, herbs and flour etc.
- The methods mentioned above are classified as dry heat or moist heat.

- Dry heat includes; broiling or grilling and roasting, which surround the meat with very hot, dry air.
- Moist heat uses lower temperature with added water, soup or sauce.
- Microwave can be used for some meats, such as bacon, sausages, hot dogs and precooked items. It is not recommended for larger cuts of meat because it cooks the meat rapidly and does not tenderize the meat.

Canning of Meat

Unlike pasteurized "cooked" meat products where the survival of heat resistant microorganisms is accepted, the aim of sterilization of meat products is the destruction of all contaminating bacteria including their spores. Heat treatment of such products must be intensive enough to inactivate/kill the most heat resistant bacterial microorganisms, which are the spores of Bacillus and Clostridium. In practice, the meat products filled in sealed containers are exposed to temperatures above 100 °C in pressure cookers. Temperatures above 100 °C, usually ranging from 110-121 °C depending on the type of product, must be reached inside the product. Products are kept for a defined period of time at temperature levels required for the sterilization, depending on type of product and size of container.

If spores are not completely inactivated in canned goods, vegetative microorganisms will grow from the spores as soon as conditions are favourable again. In the case of heat treated processed meat, favourable conditions will exist when the heat treatment is completed and the products are stored under ambient temperatures. The surviving microorganisms can either spoil preserved meat products or produce toxins which cause food poisoning of consumers.

Amongst the two groups of spore producing microorganisms, Clostridium is more heat resistant than Bacillus. Temperatures of 110 °C will kill most Bacillus spores within a short time. In the case of Clostridium temperatures of up to 121 °C are needed to kill the spores within a relatively short time. The above sterilization temperatures are needed for short-term inactivation (within a few seconds) of spores of Bacillus or Clostridium. These spores can also be killed at slightly lower temperatures, but longer heat treatment periods must be applied in such cases to arrive at the same summary effect of heat treatment.

From the microbial point of view, it would be ideal to employ very intensive heat treatment which would eliminate the risk of any surviving microorganisms. However, most canned meat products cannot be submitted to such intensive heat stress without suffering.

- Degradation of their sensory quality such as very soft texture, jelly and fat separation, discolouration, undesirable heat treatment taste.
- Loss of nutritional value (destruction of vitamins and protein components).

In order to comply with above aspects, a compromise has to be reached in order to keep the heat sterilization intensive enough for the microbiological safety of the products and as moderate as possible for product quality reasons.

A method was developed for such a balance between food safety and food quality requirements by measuring and quantifying the summary amount of heat treatment to which a canned product is exposed during the entire sterilization process.

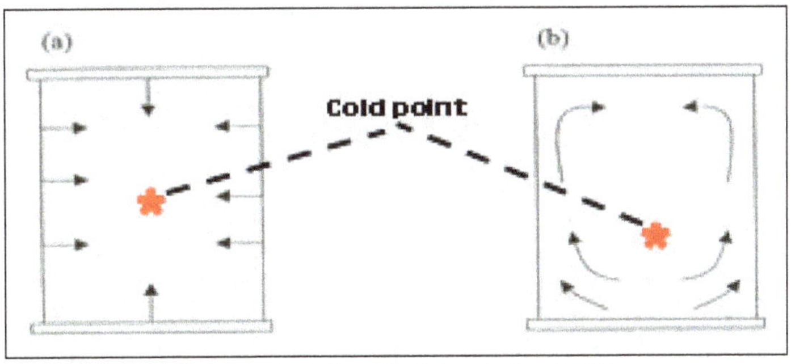

Cold point in cans with solid (a) and liquid (b) content.

The amount of heat treatment applied to a meat product can be measured using the F-value-concept. This concept is practiced in canning plants, in particular as part of the HACCP-system. Small producers, who are not equipped to employ the F-value concept, are not excluded from producing properly sterilized canned goods. In these cases established technical reference parameters of sterilization temperatures and times to the type of product and to the size of the cans can be employed. The size and format of cans is of utmost importance for the speed of heat penetration. Temperatures to be achieved at the "cold point" of the can where the heat arrives last are reached faster in small cans due to the shorter distance to the heat source than in large cans.

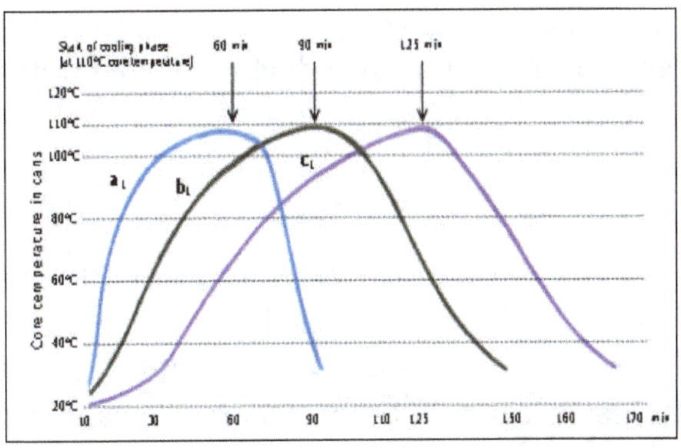

Heat penetration into centre of cans (Cans filled with meat mix of Luncheon meat type). Temperature of autoclave water 118 °C.

When comparing cans with the same bottom area (99 mm) but different height (36, 63 and 113 mm), heat penetration to the cold point of the high can takes twice the time as needed for the lowest can (a_1). These are approximate values, which can differ slightly depending on the materials filled into the cans. When comparing cans of same volume but different format, heat penetrates faster to the cold point of flat cans (a_2, b_2) than to compact square cans (c_2), although the content (volume) of all these cans is the same.

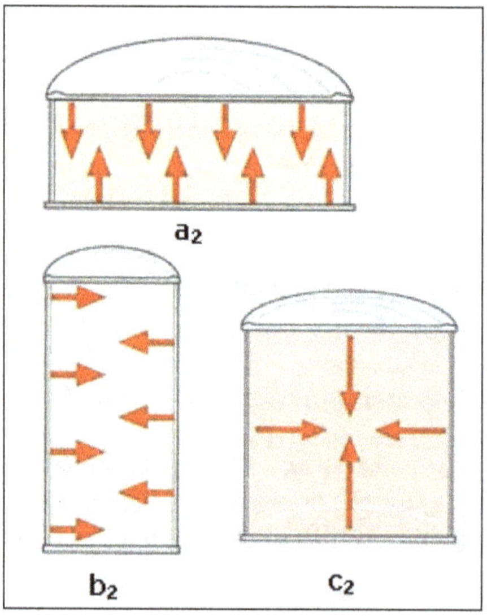

Ways of heat penetration into horizontal (a_2) and vertical (b_2) flat cans and square (c_2) cans with solid (not liquid) content.

Processes and Equipment

Process

The sterilization process in the canned product can be subdivided into three phases. By means of a heating medium (water or steam) the product temperature is increased from ambient to the required sterilization temperature (phase 1 = heating phase). This temperature is maintained for a defined time (phase 2 = holding phasing). In (phase 3 = cooling phase) the temperature in the can is decreased by introduction of cold water into the autoclave.

Autoclaves or Retorts

In order to reach temperatures above 100 °C ("sterilization"), the thermal treatment has to be performed under pressure in pressure cookers, also called autoclaves or retorts. In autoclaves (retorts) high temperatures are generated either by direct steam injection, by heating water up to temperatures over 100 °C or by combined steam and water heating. The autoclave must be fitted with a thermometer, pressure gauge, pressure relief valve, vent to manually release pressure, safety relief valve where steam is released when reaching a certain pressure (e.g. 2, 5 bar), water supply valve and a steam supply valve. The steam supply valve is applicable when the autoclave is run with steam as the sterilization medium or when steam is used for heating up the sterilization medium water.

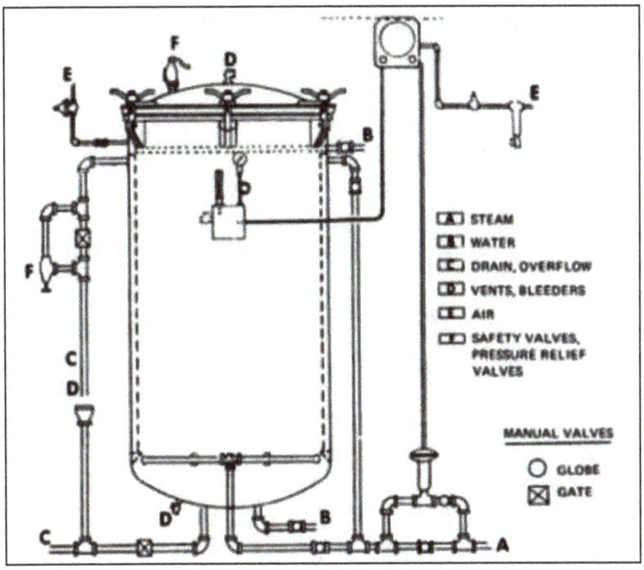

Vertical retort.

Simple small autoclaves are usually vertical autoclaves with the lid on top. Through the opened lid the goods to be sterilized are loaded into the autoclave. The cans are normally placed in metal baskets. The baskets are placed in the autoclave, either singly or several stapled on top of each other. Before starting the sterilization, the lid must be firmly locked onto the body of the autoclave. The autoclave and lid are designed to withstand pressures up to 5.0 bars. These types of autoclaves are best suited for smaller operations as they do not require complicated supply lines and should be available at affordable prices.

Double purpose equipment (for small-scale). Can be used as cooking vat or autoclave.

Larger autoclaves are usually horizontal and loaded through a front lid. Horizontal autoclaves can be built as single or double vessel system. The double vessel systems have the advantage that the water is heated up in the upper vessel to the sterilization temperature and released into the lower (processing) vessel, when it is loaded and hermetically closed. Using the two–vessel system, the heat treatment can begin immediately without lengthy heating up of the processing vessel and the hot water can be recycled afterwards for immediate use in the following sterilization cycle.

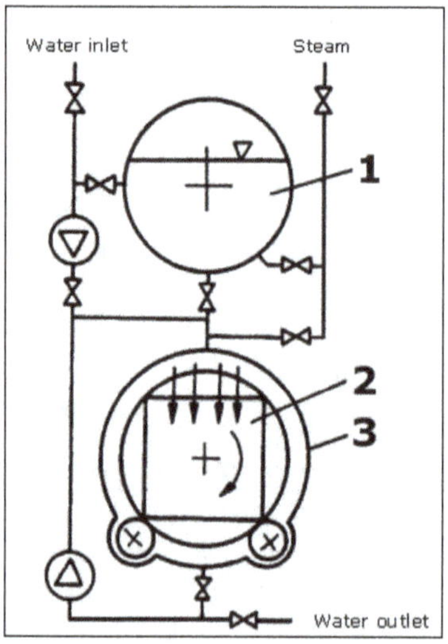

Double vessel system: 1 = Upper vessel, 2 = Basket (for rotation), 3 = Lower vessel.

If steam is used instead of water as the sterilization medium, the injection of steam into a single vessel autoclave will instantly build up the autoclave temperature desired for the process. Another technology employed is rotary autoclaves in which the basket containing the cans rotates during sterilization. This technique is useful for cans with liquid or semi-liquid content as it achieves a mixing effect of the liquid/semi-liquid goods resulting in accelerated heat penetration. The sterilization process can be kept shorter and better sensory quality of the goods is ensured.

Large horizontal autoclave.

At the final stage of the sterilization process the products must be cooled down as quickly as possible. This operation is done in the autoclave by introducing cold water. The contact of cold water with steam causes the latter to condense with a rapid pressure drop in the retort. However, the overpressure built up during thermal treatment within the cans, jars or pouches remains for a certain period. During this phase, when the outside pressure is low but the pressure inside the containers is still high due to high temperatures there, the pressure difference may induce permanent deformation of the containers.

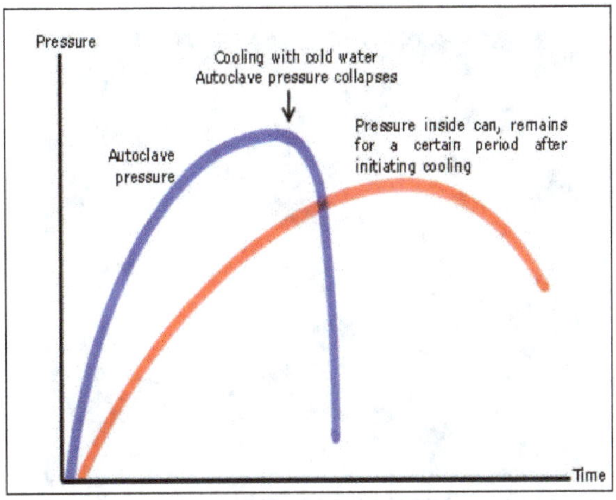

Pressure inside autoclave (blue) and compressed air inside cans (red) during heating and cooling phase (schematic).

Therefore, high pressure difference between the autoclave and the thermal pressure in the containers must be avoided. This is generally achieved by a blast of compressed air into the autoclave at the initial phase of the cooling. Sufficient hydrostatic pressure of the introduced cooling water can also build up counter pressure so that in specific cases, in particular where strong resistant metallic cans are used, the water pressure can be sufficient and compressed air may not be needed. For the stabilization of metallic cans, stabilization rims can be moulded in lids, bottom and bodies.

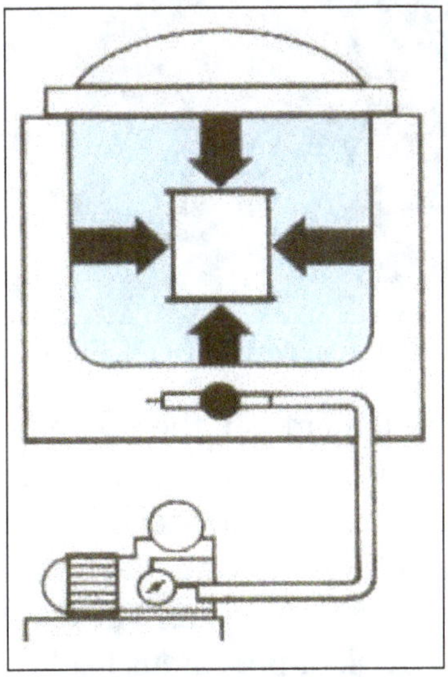

Producing counter pressure on cans inside the autoclave with compressed air.

Types of Containers for Thermally Treated Preserves

Containers for heat-preserved food must be hermetically sealed and airtight to avoid recontamination from environmental microflora. Most of the thermally preserved products are in metal containers (cans). Others are packed in glass jars or plastic or aluminum/plastic laminated pouches.

Three pieces steel cans of different format, inside lining with synthetic compound, side wall induction soldered, lid with easy-to-open ring.

Most metal containers are cans or "tins" produced from tinplate. They are usually cylindrical. However, other shapes such as rectangular or pear-shaped cans also exist. Tinplate consists of steel plate which is electrolytically coated with tin on both sides. The steel body is usually 0.22 to 0.28mm in thickness. The tin layer is very thin (from 0.38 to 3.08 µm). In addition, the interior of the cans is lined with a synthetic compound to prevent any chemical reaction of the tinplate with the enclosed food.

Cans of different size but same format (= relation of diameter to height).

Tin cans consist of two or three elements. In the case of three-piece steel cans, they are composed of the body and two ends (bottom and lid). The body is made of a thin steel strip, the smaller ends of which are soldered together to a cylindrical shape. Modern cans are induction soldered and the soldering area is covered inside with a side-strip coating for protection and coverage of the seam. The use of lead soldered food cans was stopped decades ago. Hence the risk of poisonous lead entering canned food no longer exists.

Tin cans and aluminium cans of different sizes and shapes.

Two-piece steel cans have a lid similar to the three-piece cans but the bottom and body consist of one piece, which is moulded from a circular flat piece of metal into a cup. These cup-shaped parts may be shallowdrawn (with short side wall) or deep-drawn (with longer side walls). However, the length of the side walls is limited through the low moulding ability of steel (example: tuna tins 42/85mm, i.e. side wall: diameter =1:2).

Two-piece cans. Left steel can, right aluminium can. Steel can be shallow drawn, aluminium can be deep drawn.

Aluminium is frequently used for smaller and easy-to-open cans. Aluminium cans are usually deep-drawn two-piece cans, i.e. the body and the bottom end are formed out of one piece and only the top end is seamed on after the filling operation. The advantages of aluminium cans compared to tin cans are their better deep-drawing capability, low weight, resistance to corrosion, good thermal conductivity and easy recyclability. They are less rigid but more expensive than steel plate cans.

Glass jars are sometimes used for meat products but are not common due to their fragility. They consist of a glass body and a metal lid. The seaming panel of the metal lid has a lining of synthetic material. Glass lids on jars are fitted by means of a rubber ring.

Glass jars.

Retortable pouches, which are containers made either of laminates of synthetic materials only or laminates of aluminium foil with synthetic materials, are of growing importance in thermal food preservation. Thermo-stabilized laminated food pouches, have a seal layer which is usually PP (polypropylene) or PP-PE (polyethylene) polymer, and the outside layers are usually made of polyester (PETP) or nylon. They can be used for frankfurters in brine, ready-to-eat meat dishes etc. From certain laminated films, for instance, polyester/polyethylene (PETP/PE) or polyamide/polyethylene (PA/PE), relatively rigid containers can be made, usually by deep drawing. They are used for pieces of cured ham or other kinds of processed meat. Small can-shaped round containers are made from aluminium foil and polyethylene (PE) or polypropylene (PP) laminate and are widely used for small portions, particularly of sausage mix. PE or PP permits the heat-sealing of the lid made of the same laminate onto these containers, which can then be subjected to intensive heat treatment of 125 °C or above. One advantage of the retortable pouches/laminated containers is their good thermal conductivity which can considerably reduce the required heat treatment time and hence is beneficial for the sensory product quality.

Cans moulded of strong aluminium foil, lid made of flexible aluminium foil to be heat sealed on body.	Cans made of synthetic material with sealed aluminium foil as lid (can be torn off).

Cleaning of Containers Prior to Filling

Rigid containers (cans, glass jars) are delivered open to meat processing plants, i.e. with the lids separate. During transport and storage, dust can settle inside the cans, which must be removed prior to filling the cans. This can be done at the small-scale level by manually washing the cans with hot water. Industrial production canning lines are equipped with steam cleaning facilities, where steam is blown into the cans prior to filling.

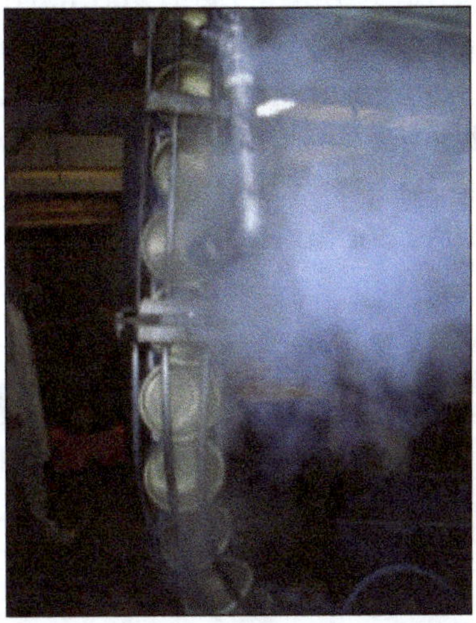

Empty cans be cleaned by steam injection.

Seaming of Cans

After the can is filled with the product mix the can is sealed with a tight mechanical structure - the so-called double seam. The double seam, in its final form and shape, consists of three layers of lid (D, black colour) and two layers of body material (D, striated). The layers must overlap significantly and all curves must be of rounded shape to avoid small cracks. Each double seam is achieved in two unit operations referred to as "first operation" (A, B) and "second operation" (C, D).

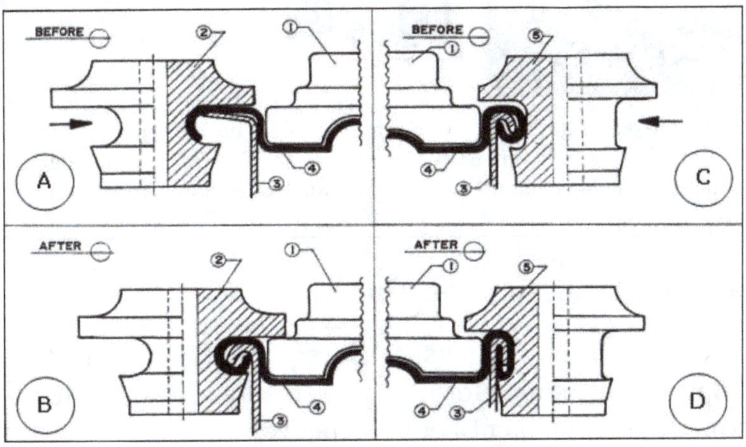

Can seaming operation (schematic).

The can covered with the lid is placed on the base plate of the can seaming machine. The can is moved upwards while the seaming chuck keeps the lid fixed in position. The pressure applied to the can from the base plate can be regulated and must be strong enough to ensure simultaneous movement of the lid and the can to avoid scratching-off of the sealing compound.

In the first operation the lid hook and body hook are interlocked by rolling the two into each other using the seaming roll with the deep and narrow groove. The body hook is now almost parallel to the lid hook and the curl of the lid adjacent to or touching the body wall of the can. In the second operation, the interlocked hooks are pressed together by a seaming roll with a flat and wide groove. Wrinkles are ironed out and the rubber-based material is equally distributed in the seam, filling all existing gaps thus resulting in a hermetically sealed container.

Design of Seaming Rolls

The seaming rolls for the first and second operations are designed differently in order to facilitate the respective operations. The seaming roll for the first operation has a deep but narrow groove to interlock body and lid hock (rolling the hocks into each other). The seaming roll for the second operation has a flat but wide groove to press the interlocked hooks together (sealing the seam). The different grooves of the first and the second seaming roll are shown in the pictures below. The first action (first roll) is rolling (interlocking) the hooks, the second action (second roll) is compressing (sealing) the seam.

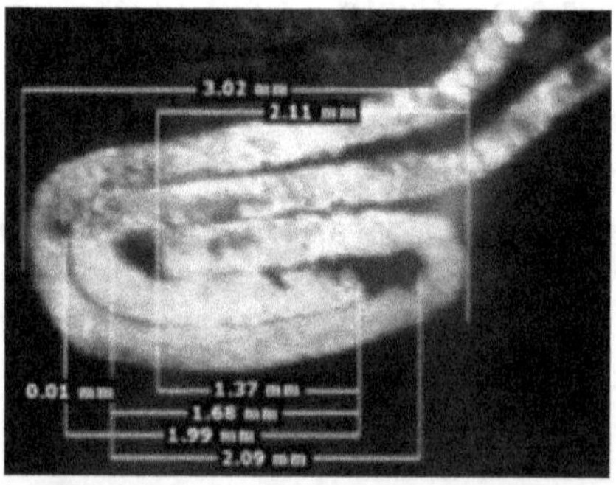

Cross section of double seam: Routine check of correct seam in cannery by cutting out seam segment and using computerized testing.

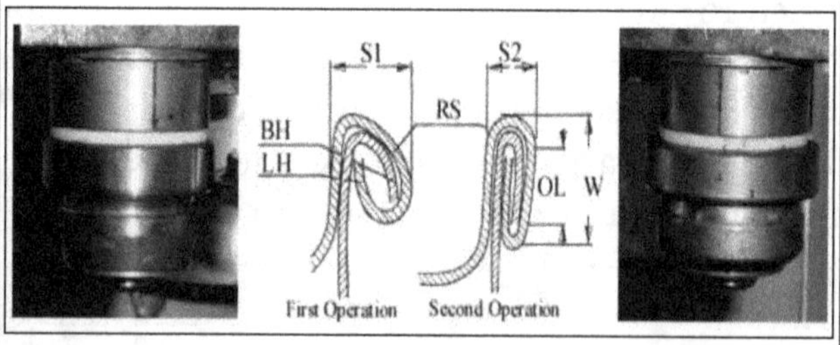

Can seaming using semi-manual can seaming machine. S1 - First operation W - Width of the seam BH - Body hook OL – Overlap. S2 - Second Operation LH - Lid hook RS - Rubber seal.

Meat Products Suitable for Canning

Practically all processed meat products which require heat treatment during preparation for consumption are suitable for heat preservation. Meat products which do not receive any form of heat treatment before being consumed, such as dried meat, raw hams or dry sausages, are naturally not suitable for canning as they are preserved by low pH and/or low water activity. The following groups of meat products are frequently manufactured as canned products:

- Cooked hams or pork shoulders.

Canned pork shoulder.

- Sausages with brine of the frankfurter type.
- Sausage mix of the bologna or liver sausage type.

Bologna sausage mix, variety of same type is Luncheon meat.

- Meat preparations such as corned beef, chopped pork.

Corned beef.

Corned pork.

- Ready-to-eat dishes with meat ingredients such as beef in gravy, chicken with rice.

Meat mix with potatoes.

- Soups with meat ingredients such as chicken soup, oxtail soup.

Pork knuckle meat in gravy ("Adobo", a Philippine specialty).

Effects of Processing and Preservation Technologies on Meat Quality

The changes produced in meat due to the application of different processing techniques, preservation methods, and technologies can be basically of two types: physical and chemical. Physical changes are modifications in the structure of the tissues that affect the sensory characteristics of the product such as volume, appearance, color, texture, aroma, and taste. Different effects in meat can be cited; reduced surface moisture due to dehydration, increased moisture and fat retention due to protein denaturation, and enhanced functional properties of proteins due to incorporated additives.

The chemical changes in meat are due to the molecular interactions that occur when thermal treatment is applied, food additives are added, or when storage is prolonged. When the chemical structures of the substances responsible for organoleptic characteristics or nutritional value are affected, for instance in the denaturation, hydrolysis, and gelation suffered by proteins due to the actions of boiling water and prolonged heating times, the consequences influence the consumer acceptance and affect balanced diet. Technologies which ensure food safety and meet the demands of the consumers without compromising the nutritional value of traditional meat products are required.

Consumers demand preservative-free, minimally processed meat products with a longer shelf life. Nowadays the use of natural additives instead of synthetic additives is being widely accepted. In addition to this, research on more eco-friendly packaging materials, which improves the shelf life of meat, is gaining momentum. The development of new meat products with improved nutritional profiles has increased over the last decade. For this purpose, there are two main strategies: obtaining healthier fresh meat and post-mortem processing of meat products. These strategies could affect the quality of meat products and their nutritional value.

The meat processing method is usually selected mainly focusing on the technological, microbiological and healthy aspects of the product. However, when selecting a processing and/or preservation technology, not only the quality impact on the product should be considered; a comprehensive and global strategy considering the changes in sensory and nutritional features and consumer appeal is necessary.

Processing of Physical Methods

Dry Aging

Dry aging is the process of ripening of meat at controlled conditions. The meat carcasses or primal cuts are hanged in a refrigerated chamber (0–4 °C) with the relative humidity maintained between 75 and 80% for 28–55 days. Until now, only bovine and porcine meats have been investigated. The process is comparably costly due to the need of quality meat cuts, shrinkage loss (6–15%), trim loss (3–4%), and the high risk of open-air contamination in meat. Open-air contamination can be reduced by packaging the meat in highly moisture permeable bags.

The effects of dry aging treatment on meat quality are summarized in table. Dry aged meat has excellent flavor and palatability as a result of proteolysis, lipolysis, and concentration of flavor compounds due to water loss. Dry ageing imparts brown-roasted, beefy, buttery, nutty, roasted-nut, and sweet flavor in bovine meat. In beef and pig, dry aged meat has an umami taste due to the high level of glutamate. In a comparative study on dry aged and vacuum aged meat, it was observed that the umami and butter fried taste were more prominent in dry aged meat. Moreover, the consumer opinion on sensory aspects of dry aged meat was better in comparison to vacuum aged meat, the meat was found to be more tender and juicier. Dry ageing improves the tenderness and juiciness of bovine and porcine meat.

Table: Effects of some physical treatments on the sensory and nutritional characteristics of meat products.

Treatment	Meat Product	Effects
Dry aging	Beef meat.	More flavor, tenderness and juiciness in beef.
	Porcine meat	
		Umami taste in beef and porcine meat. Nutritional changes not investigated.
Dry curing	Pork, beef, mutton	Increased storage temperature. Slightly decreased the digestibility of dried pork protein.
	Meat products	Protein quality is not significantly reduced during dehydration.
	Meat products from different animals	Hardened texture, wrinkled appearance, characteristic flavor, brown color and darkening.
High pressure processing	Beef, pig, chicken meat	Unchanged nutritional value.
	Different meat products	Low molecular weight vitamins and flavor compounds stay intact.
	Beef and rabbit muscle	Enhanced digestibility.
	Meat products	Improved tenderness, changes to the color quality depending on the content of myoglobin.
	Goose breast	Improved tenderness.
	Lamb meat cuts	Browned, livery and oxidized flavors.
	Ham	Improved digestibility.
	Ready to eat (RTE) meat products	No changes in sensory properties.
	Pig ham	Paler color and softer texture.
Low-temperature long-time (LTLT) and sous vide cooking	Meat	Increased tenderness and better appearance.
	Lamb and pork meats	Increased flavor.
	Beef	Increased tenderness.
	Chicken meat	Increased tenderness and color.
	Beef	Brownish red with a slight green color.
	Pork	Juicier and tenderer meat and improved digestibility.
	RTE marinated beef	No effects on sensory characteristics.
3D printing	Turkey and beef meats	Novel appearance and texture. Nutritional changes not investigated.

Dry Curing

Dehydration is the process of reducing the moisture content in meat to improve its shelf life. Automated drying chambers with programmable logic controller and real time monitoring are nowadays widely used in the meat industry. In these chambers the air-flow rate, temperature, relative humidity, and flow distribution can be controlled relative to the size, shape, structure, and moisture content of the product. The water holding capacity, the state of muscle proteins and its microscopic structure determines the rehydration property of the dehydrated meat. The muscle fiber diameter as well as the space between the groups of muscle fibers reduces during dehydration. The rate of reduction in the moisture content during dehydration is high in precooked meat compared to raw meat.

The heat damage during dehydration of meat is characterized by the burnt flavor, toughness, and grittiness. The conceptualization of the distribution of water in meat during dehydration can help optimize the process, which can be done by novel non-destructive techniques like hyperspectral imaging. Researchers have used the technique effectively in beef slices where the pixel wise images were taken at different time periods at six specific wavelengths. Regarding the nutritional value of "dehydrated meat", only two studies were carried out, and they were in the 1940s. They referred to dehydrated and packaged meat obtained with methods now in disuse. Most countries have their own traditional dried meat products, which have similar sensory features. Kilishi is a traditional sun-dried meat product, which is spiced and roasted, with a shelf life of around 12 months. Biltong is another product where meat is salted and dried. Both products are commonly consumed in African countries. Carne do sol and charque are traditional salted dried Brazilian meat products. Tasajo, sou gan, pastirma, and cecina are salted and dried meat products traditionally prepared and consumed in regions like Cuba, China, East Mediterranean, and Mexico/Spain, respectively. Bresaola, jamón serrano, sucuk are traditional salted fermented and dried meat commonly consumed in Italy, Spain, and Turkey.

Dried meat products have a hardened texture and wrinkled appearance due to the volume reduction, and sometimes the meat has a hard crust on the surface. Aroma compounds are produced in the meat products as a result of lipid oxidation that imparts a characteristic flavor to the meat. The dried meat has a brown color, the color changes from red to brown according to the temperature. The salt added during drying also adds to the darkening effect. Nitrate/nitrites can also be added to modify the color and flavor of the meat. In dry cured products, the characteristic flavor is due to the metabolites produced as a result of the action of enzymes on meat.

High Pressure Processing

High pressure processing (HPP) is a non-thermal decontamination minimal processing technology, where the meat is subjected to a pressure range of 350–600 MPa for a few minutes to acquire improved microbiological safety and shelf life. Pressure is exerted isostatically and the volume of the product decreases with the increase in pressure.

HPP can affect the sensory and nutritional characteristics of meat products. Application of high pressure breaks the less strong ionic bonds and hydrogen bonds, which in turn denatures the protein via alteration of the quaternary structure of protein followed by tertiary structure at higher pressure ranges. The nutritive value of the meat is minimally affected by HPP. The low molecular weight vitamins and flavor compounds stay intact since pressure does not affect covalent bonds. High pressure treatment can potentially be an effective technology to improve the digestibility of meat products. This effect has been more pronounced in the muscles treated at 600 MPa.

A pressure higher than 200 MPa leads to the changes in meat protein with various effects of gelation, aggregation, and changes in texture due to the making and breaking of bonds. The effects also vary according to the range of pressure applied and the time of application of the pressure. Meat subjected to high pressure tends to conform to a gel consistency when the secondary and tertiary structure of protein breaks down keeping the primary structure intact. The characteristic structure of myoglobin changes with the application of pressure and it forms a new aggregated protein conformation with reduced solubility. The elasticity of meat increases making it more tender. HPP tends to modify the texture of the meat by tenderizing it, since high pressure fractures rod-like muscles in meat. HPP induces unfolding of myofibrillar proteins, which subsequently exposes sulfhydryl and hydrophobic groups to the surface, unraveling helical structures and forming myosin oligomers through disulfide bond.

The meat treated with HPP at high pressure levels of 400 and 600 MPa were associated with browned, livery, and oxidized flavors, which will have an impact on the consumer and market behavior of the product. There is no immediate effect of HPP on the oxidative stress of meat. Meat processed at higher pressures for longer period is tougher than the meat processed at a lesser pressure for lesser time. The intrinsic properties of HPP processed meat differs according to the processing conditions applied and the type of raw material. HPP tends to induce aggregation, which improves the digestion of the meat. Some studies did not observe any significant difference in the sensory properties of high pressure processed ready to eat (RTE) meat. When smoked pork rounds were subjected to a pressure of 600 MPa for three minutes significant differences were observed in the cohesiveness and the odor of the meat, whereas the other textural and sensory properties were not affected by the pressure treatment.

The HPP treated samples of ham had paler color and softer texture compared to normal ham samples. Some studies concluded that HPP at 500 MPa combined with mild heat treatment at 53 °C was optimal for production of ham. When goose meat was subjected to HPP at an optimal condition of 213 MPa for 15 min, it was observed that pressure range and the time of holding significantly affected the hardness, since the rod like muscles were fractured. High pressure does not have much effect on cooking loss rate or water holding capacity. HPP at 450 MPa and 600 MPa did not significantly change the properties of seared beefsteaks in term of pH, water activity, moisture content, and expressible moisture. However, enhanced water holding capacity was observed in rabbit muscles when subjected to HPP.

Conventional Cooking

Cooking makes foods safe to consume and palatable. In order to guarantee food safety, food is cooked at higher temperatures for longer time; however, this practice decreases the nutritional and organoleptic quality of the foods; loss and oxidation of water soluble and thermolabile vitamins, loss of fats due to fusion, chemical browning reactions, etc.

The effects of cooking temperatures on proteins are varied. At temperatures up to 100 °C, as occurs in water or microwave cooking, this denaturation translates into effects of interest, such as enzymatic inactivation of lipases, proteases, etc., improvement of digestibility or reduction of toxicity; between 100 and 140 °C, as in pressure cooking and baking, digestibility is reduced by forming intramolecular and intermolecular covalent bonds. The same effects happen at temperatures above 140 °C, as in frying and roasting on the grill, where amino acid destruction occurs, such as cysteine or tryptophan, with isomerization to D-configuration and reduction of nutritional value. In lipids, heat treatment produces fusion, although being triglyceride mixtures it is difficult to establish its exact melting point; before reaching the liquid state, they go through a pasty state, then smoky (at a different temperature depending on the type of fat) and then decompose. Even intense heating can sometimes form toxic cyclic monomers, dimers and polymers, as is the case with acroleins. Carbohydrates are generally considered stable against cooking. However, solubilization losses of these compounds, which depend on factors such as time, size, etc., cannot be avoided.

With traditional cooking systems, there is a long waiting time between the preparation and the distribution of meals. Therefore, food must be placed in hot cabinets, ovens, water baths, etc., to avoid its cooling, where the food is dried and over-cooked. The result is a lukewarm meal, with a temperature below 65 °C in the center of the product, and therefore hygienically dangerous as these storage temperatures allow the growth of mesophilic and thermophilic microorganisms that will contaminate dishes during the waiting time for service and consumption. This fact can be dangerous in places of collective catering, such as hospitals, nursing homes, and school canteens, where the group to which the menu is directed has compromised immune system.

New Techniques of Cooking: Low-Temperature Long-Time (LTLT) and Sous Vide Cooking

LTLT cooking has numerous advantages, of which the most sought-after characteristics are controlled doneness, improved tenderness and uniform eating quality. In LTLT cooking the product reaches a thermal equilibrium with the medium of heating, which contributes to these additional advantages of the product over traditional high temperature cooking. The underlying mechanism which provides more tender meat, regardless of the age of animal, species, or type of muscle, at an optimum combination of temperature and time has not yet been completely elucidated. It might be, possibly, due to the interaction between proteolysis of myofibril structures and heat induced denaturation of proteins. The reduction of LTLT cooking temperature and holding time improve the juiciness of the meat, but at the same time in a constricted temperature range, higher cooking time imparts the desired aroma and flavor characteristics to the cooked meat. The flavor intensity of the LTLT cooked meat is medium to low in comparison with the meat cooked at higher temperature. The long cooking time weakens the forces holding the myofibrils together in aged meat leading to meat fragmentation upon shearing, and in meat with less amount of connective tissue the degree of tenderization is relatively high when cooked at 50–60 °C. Prolonged heating time denatures the protein even if the temperature of cooking is lower than the actual temperature of denaturation.

Sous vide cooking (vacuum cooking) is new variant cooking technique used normally to produce high-quality dishes in the food service sector. Food is vacuum packed in a heat-stable plastic pouch, followed by incubation in a water bath at controlled conditions of time and low temperatures (53–81 °C). The cooking temperature is maintained lower with a higher cooking time. This technique maintains a uniform meat quality and improves the organoleptic property of the cooked meat. Sous vide cooked meat is more tender and redder than conventionally cooked meat. The duration and the temperature of the cooking comparably affect the physicochemical characteristics and palatability of meat.

Some studies have shown that there is significant effect of the cooking time and temperature over the texture of the meat. In sous vide cooked meat the increase of the cooking temperature and time result in increased shear force and toughening, respectively. However, the shear force is reduced when sous vide is combined with other treatments. Water loss in the meat results in shrinkage of the muscular fibers both transversally and longitudinally, aggregation and gelling of sarcoplasmic proteins, shrinkage and solubilization of connective tissues, which leads to the formation of granular fibers. If the sous vide cooking is carried out at higher temperatures the cooking loss is maximum with a minimal reheating loss, due to the increased shrinkage caused by denaturation of the proteins. In some studies of sous vide cooking, an increase of the opacity of meat surface was observed, which was due to the water loss. In sous vide cooked meat the reddish color of meat is replaced by a brownish red with a slight green color since the deoxymyoglobin and oxymyoglobin is denatured with an increase in the metmyoglobin and sulfmyoglobin as a result of longer cooking time. The shelf life of sous vide cooked chicken tikka masala, a traditional Indian meat delicacy was comparatively high (40 days) with slight change in color. The higher shelf life was due to the spices and herbs in the product.

Sous vide processing renders meat juicier and tenderer and at the same time, the technique improves the digestibility of meat, according to studies conducted on in vitro digestion. However, in a study with young men, no differences were observed between the digestibility of sous vide cooked and fried meat in the pan. Digestibility remains unknown in elderly adults. The volatile profile of sous vide cooked meat is better preserved with little accumulation of off flavor imparting compounds such as hexanal or

3-octanone usually found in traditionally cooked meat. A higher retention of vitamin B_3 is another advantage of sous vides cooking since the cooking temperature is retained at a comparatively lower level.

Sous vide cooking can be carried out as low temperature long time or high temperature short time treatments. When the LTLT sous vide cooking method is used, collagen solubilizes, and a larger amount of gelatin is formed with less intense myofibrillar toughening. The high temperature short time alternative can be considered as more economic and feasible method due to the higher safety and comparable quality attributes of the cooked meat, but with a lesser retention of vitamins and higher hardness.

Sous vide technique can be combined with other techniques, such as marination. For instance, reported the feasibility of using the combination of marination and sous vide cooking techniques to yield new RTE meat products with high protein content and without negative characteristics. In this way, the benefits of two different techniques are taken advantage without compromising the quality of the product.

3D Printing

3D printing is a novel pre-processing technology used in foods, where extrudable food is printed into specific shapes of uniform structure or layers using a 3D printer. There are several categories and ingredients of printable food. 3D printing enables automation, waste reduction, and personalization of foods. Meat should be processed into an extrudable form with added binders or texturizers, such as hydrocolloids or gellable proteins, so that meat can be 3D printed. In addition to this, meat should be in a viscoelastic form so that it can be printed into the specific structure.

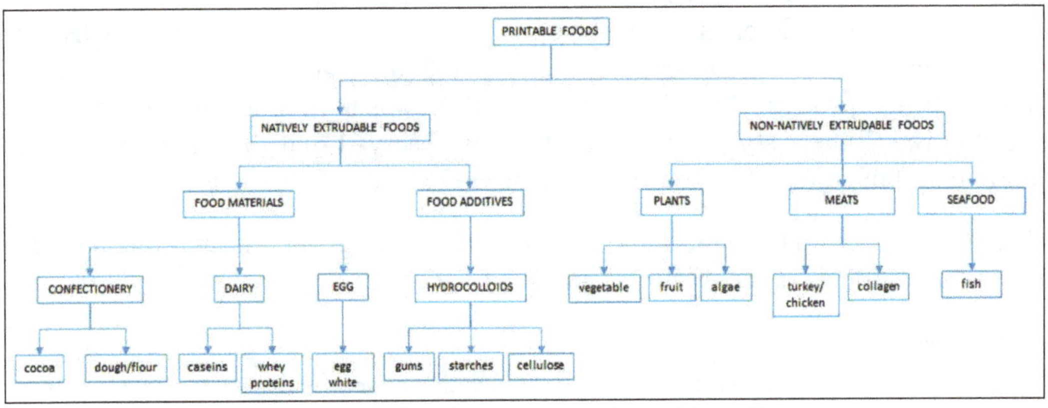

Printable food classification, categories and ingredients.

Few studies have studied the effects of 3D printing on sensory and nutritional characteristics of meat. Turkey puree with added binders and viscosity enhancer was successfully printed for sous vide cooking. Researchers have conducted successful studies on the use of fibrous meat for 3D printing in medical field for elderly and patients who require ketogenic diet. Moreover, beef paste prepared with guar gum binder and lard was 3D printed into multiple layers and sous vide cooked. The cooked samples maintained the structure with slight inward contraction in all the layers. It was observed that increasing the lard layers led to higher cooking loss, shrinkage, cohesiveness, lower fat retention, moisture retention, hardness, and chewiness, whereas increasing the infill density led to higher moisture retention with lower shrinkage and cohesiveness, resulting in higher hardness and chewiness.

Processing of Chemical and Biochemical Methods

Fermentation

Fermented meat products are mainly dry cured sausages, commonly eaten in many regions worldwide. In the Mediterranean regions, they are medium humidity meat products with considerable shelf life elaborated with spices such as paprika, garlic, and black pepper filled in casings and further cured or ripened so that the flavor would be enhanced. Prior nitrite treatment is considered as a mandatory pre-treatment in most of the European countries.

The nitrogen compounds in the meat muscles are denatured enzymatically imparting meat the characteristic flavor. The enzymes like protease, aminopeptidases, and microbial enzymes breaks down the proteins in muscles, generating small peptides and amino acids, like alanine, leucine, valine, arginine, lysine, glutamic and aspartic acids, which imparts meat the characteristic flavor. In some cases, the stage of curing is assessed based on the concentration of these amino acids.

The effects of fermentation on the nutritional and sensory characteristics of meat are listed in Table. The flavor and quality of the finished product depends on the process duration. The color is determined by the amount of sarcoplasmic protein. The pH of the product decreases during the fermentation, leading to the gelation of the sarcoplasmic and myofibrillar proteins. Lactobacillus fermentum was used as a substitute for nitrite in Harbin red Chinese style sausage, and it was observed that the characteristic pink color of cured meat was retained in the fermented meat.

Table: Effects of fermentation, smoking, curing and salting, and marination on the sensory and nutritional characteristics of meat products.

Treatment	Meat Product	Effects
Fermentation	Dry-cured meat products, traditional Jinhua ham, Parma ham, dry-cured Iberian ham	Specific aroma compounds such as alcohols, aldehydes, ketones, esters and lactones.
	Harbin red Chinese style sausage	The use of Lactobacillus fermentum led to characteristic pink color of cured meat.
	Fermented meat products	Production of bioactive peptides.

Smoking	Lamb meat	Enhanced flavor, color and odor.
	Sausage	Reduction of the greyish discoloration.
	Sausages from poultry, pork and beef meat	Enhanced sensory attributes.
	Smoked pork bacon	Improvement of digestibility of indispensable amino acids.
	Buffalo meat	Combination of smoking with curing improved color, texture and odor.
Curing and salting	Bovine muscle	Improved texture properties.
Marination	Broiler chicken Fresh and precooked pork meat	Increase of the juiciness and tenderness.
	Beef	Tripolyphosphate in brine improved sensory characteristics.
	Beef	Brines of salts and phosphate improved juiciness and tenderness. Color loss.
	Beef	Acid concentrations greater than 0.3 M were not recommended, as they caused great swelling and darkening.
	Beef	The higher acid concentration used for the brine, the greater tenderness. Solution with an acid concentration greater than 0.15 M leads to too acidic beef and rejection by panelists.

The secondary oxidation products, formed as a part lipolysis and auto-oxidation in the lipids, develop specific aroma compounds like alcohols, aldehydes, ketones, esters, and lactones during the fermentation of the meat. It has been found that meat proteins can produce bioactive peptides making it more susceptible to be used as functional ingredient too.

Presently in the food industry, starter and protective cultures are used, rather than relying on the natural microflora to ensure the sensory and microbial quality of the fermented meat products. During fermentation it has been found that various bacteriocins are produced, which inhibits the growth of other spoilage and pathogenic microorganism.

Smoking

Smoking is an age-old preservation technique, where meat is subjected to smoke, which affects the sensory and nutritional characteristics of meat products. There are positive effects, such as improvement of flavor, color and odor in lamb meat. The effect of smoking on meat increases with the time of exposure. Hot smoking, cold smoking, electrostatic smoking and use of condensates, smoke aromas or liquid smoke are different kinds of smoking treatments. Meat is smoked at 20–25 °C at a relative humidity of 70–80% and at 75–80 °C during cold and hot smoking, respectively. The electrically charged smoke particles which precipitate over the meat in electrostatic smoking reduces the time of processing. In products where the protein denaturation which accompanies the smoking process is deemed undesirable, the smoke aromas or condensates are used.

Smoke process is an effective treatment against pathogenic microorganisms (Staphylococcus aureus, Escherichia coli, Listeria monocytogenes, Salmonella spp., etc.), and reduces the lipid oxidation, which leads to undesirable flavors and oxidative rancidity. In sausages smoking helps to reduce the greyish discoloration. Smoking allows incorporating different specie meats to obtain high quality sensory sausages. The sensory scores for smoked buffalo rumen meat products added with ginger extract were found to be within the acceptable limit during the storage period of 15 days. Meat smoking enhances the sensory attributes but at the same time contaminates with carcinogenic residues such as polycyclic aromatic hydrocarbons or nitrosamines. The residues are nowadays reduced by separating the smoke generating chamber and the smoking chamber, such that the residues precipitate in the generation chamber, exempting the meat of theses harmful residues.

Smoking reduces the water activity of meat, which affects the hardness of the product and the protein stability. The digestible indispensable amino acids, which help in evaluating the protein quality in food, were calculated in smoked bacon and improved the content after smoking. Curing combined with smoking has been found to increase the pH and improve the color, texture and odor of buffalo meat. Smoke curing has the combined effect of both enzyme and heat, which leads to alterations in the fatty acid profile of pork and lamb meat.

Curing and Salting

The main aim of adding salt into meat was to preserve the meat, but now the cured meats have a high demand due to their characteristic flavor and organoleptic properties. Nitrate and nitrite salts have been known to create the pinkish red color and characteristic flavor of meat and increase their shelf life. Nitrite salts inhibits the lipid oxidation that imparts a rancid flavor to meat. Recent studies have found a considerable amount of carcinogenic by-products formed as a result of adding nitrite to meat leading to the reduction in its use for curing opening a window for the use of organically cured meats, where nitrate of natural origin from vegetable sources are used. Smoking has been used to improve the preservative effect of meat during curing but nowadays due to the flavor imparted to meat, smoking has developed consumer appeal.

Sodium chloride plays multifunctional role in preservation and processing of meat. It increases the shelf life of cured meat by reducing the water activity of meat, which in turn reduces the microbial load. Salt plays a critical function in determining the gelation, emulsifying and linkage properties in meat muscle proteins.

Scientific organizations lately propose the decrease of salt content in processed foods, which has led to a lot of studies on meat with salt replacements, such that the palatability and texture of cured meat are not compromised. Studies on Pamplona chorizo with low salt found that the product was acceptable. The use of potassium chloride, flavor enhancers like carboxymethyl cellulose and carrageenan in combination with sodium citrate or combination of sodium, potassium and magnesium salts, or undissolved salt crystals were used in various studies to reduce the sodium chloride content in meat with partial success, since low salt meat does not have the same palatability as the meat with normal sodium chloride content.

Marination

Marination is a meat tenderization procedure, with the use chemical methods. This treatment increases the rate of natural proteolysis in meat, by means of a greater drop in the pH of the meat after slaughter, stimulating the enzymatic proteolytic activity during muscle maturation. Meat is treated with mixtures of different common organic acids like citric acid, acetic acid, and tartaric acid, from orange juice, apple cider vinegar, and agraz-verjus wine (unpublished results). It accelerates the maturation time of the meat by reducing the time necessary for its softening. The mechanism by which marinade influences meat tenderization appears to involve several factors including weakening of structures due to meat swelling, an increase in proteolysis caused by cathepsins, and an increase in the conversion of collagen to gelatin at a low pH during cooking.

When a piece of meat is directly immersed in an aqueous solution with various ingredients such as salt, organic acids, etc., the ingredients gradually penetrate by osmosis. The amount of salt and other ingredients from the peripheral parts of the final piece is superior to that of the central zones, not obtaining a homogeneous result. That is why, at an industrial level, marination by immersion is replaced by marinade injection methods, which has several effects on the sensory properties of meat on or marinade injection technology in chicken and pork has been developed for years.

The poultry industry has used water injection and polyphosphates for more than 20 years; mainly with the aim of facilitating water retention during maturation and subsequent cooking, which leads to an increase in the juiciness of the meat and, with it, an increase in the perception of tenderness by the consumer. Researchers studied the acceptability and shelf life of fresh and precooked pork meat injected with salt, dextrose, citric acid, tripolyphosphate and sodium pyrophosphate, finding that, while citric acid and pyrophosphate lowers the pH of meat, tripolyphosphate increases it, causing a decrease in microbial growth and an improvement in sensory characteristics. Researchers also used solutions of salts and phosphate in different cuts of beef, observing an improvement in juiciness and tenderness. Injection at different pressures

(345 and 200 kPa) implies differences in losses during cooking and in Warner-Bratzler shear force (WBSF). However, the softening and increase or enhancement of flavor by immersing the meat in a solution of different tenderizers or flavorings (marinated) has not been widely used in cattle; therefore, its real effectiveness is little known. Different researchers have carried out brine injection tests containing, for example, sodium chloride, sodium tripolyphosphate and sodium lactate, finding an increase in juiciness over the control that had not been injected, but without finding significant differences depending on the proportion of injected brine with respect to the initial weight of the meat. Other trials have focused on trying to alleviate some adverse effects observed as a result of brine injection, such as color loss or decreased shelf life.

Some researchers have studied the effect of marinating beef with acidic aqueous solutions. For example, researchers investigated the softening of very fine beef cuts (40 × 35 × 5 mm), obtained from muscles with high connective tissue, from carcasses kept at least 48 h in refrigeration after the slaughter of the animal, and stored under vacuum at −20 °C until use. With these cuts, they carried out immersion tests, for 20 h, in acid solutions prepared from acetic, citric or lactic acids, each one individually, or with mixtures of citrus juices (diluted orange and lemon). The results obtained led them to conclude that acid concentrations greater than 0.3 M were not recommended, as they caused excessive swelling of the meat, as well as its darkening and gelatinization.

Researchers also verified the effect of marinating pieces of neck meat, about 200 g, obtained from beef carcasses that had been maturing four days after the animal was slaughtered. In this case, the marination solutions were prepared with acetic and lactic acids, in concentrations from 0.05 M to 0.25 M. Marination was carried out for two or nine days. Data on the shear resistance force, obtained with a Warner-Bratzler method, indicated that tenderness increased slightly at two days and slightly more at nine days, mainly due to proteolysis. In general, the higher acid concentration used for the marinade, the greater tenderness measured with the Warner-Bratzler method and the greater pH decrease in meat. However, tests carried out with panelists on the same meat indicated that marinating with a solution with an acid concentration greater than 0.15 M was not recommended, since the sensory panelists found the taste excessively acidic and rejected it. Meat with pH values lower than 5.0 was acceptable only up to a point.

Reformulation

Basically two methods are possible to reformulate meat products: the elimination or reduction of components considered harmful to health (fat, saturated fatty acids, salt, nitrites, etc.) and the incorporation or increase of the content of substances with nutritional properties (dietary fiber, proteins of high nutritional value, polyunsaturated fatty acids (PUFA), monounsaturated fatty acids (MUFA), etc.). Table summarizes different examples in which different reformulation alternatives and the consequences on the sensory and nutritional characteristics of foods have been studied.

Table: Effects of the reformulation of meat products on their sensory and nutritional characteristics.

Compound	Reformulation Objective	Treatment	Meat Product	Effects
Salt	Reduction	Lowering from 2.8% to 0.5%	Hotdog sausages, bacon, ham and salami	Paler, softer, and less juicy products per low of 1.3–1.7% NaCl.
			Pork sausages	Difficulty reducing the dietary salt intake (<1.4%) without affecting acceptance.
			Chicken breasts: Reconstructed ham	
	Partial substitution	Use of spice mixes, KCl or other salts	Cooked ham	Sensory quality and general acceptability were not modified if replacement ranged 30–35%.
			Bovine and chicken meat	
			Fermented sausages	Reduced acceptance of aroma, flavor, juiciness and overall quality if NaCl was lower than 1.3%.
			Frankfurt sausages	
Fat	Reduction	Addiction of vegetable oils	Pork sausages	Darker, harder, less juicy and less flavor intensity.
			Frankfurt sausage	Better nutritional value (reduction in fat and cholesterol and increase in polyunsaturated fatty acids (PUFA) or monounsaturated fatty acids (MUFA).
			Pork sausages	
			Beef and pork sausages	
			Beef patty	
	Substitution	Replacing by vegetable or fish oils, soybean proteins, carbohydrates, and synthetic compounds	Sausages, cooked minced meat	Decrease of meat aroma and flavor intensity.
			Veal sausages	Better nutritional value (reduction in fat and cholesterol and increase in PUFA or MUFA).
			Bologne sausages	
			Beef patty	
			Spanish salami	
			Sausage	

	Enhanced nutritional value	Raw material with a high level of mono and polyunsaturated fatty acids from pigs fed with different diets	Dry fermented sausage. salchichon	The color was slightly affected.
				Improved nutritional value.
		Grass-fed or flaxseed-containing concentrates	Beef	Improved fatty acid profile by increasing content in conjugated linoleic acid (CLA), eicosapentaenoic acid (EPA) and docosahexaenoic acid DHA.
		Feeding with linseed seeds and CLA	Beef patty	No significant change in color and odor of hamburgers enriched in n-3 and CLA.
				Enhanced lipid profile.
Dietary fiber	Addition of dietary fiber	Addition of dietary fiber (inulin, rice fiber, citrus fiber, etc).	Sausages	Texture properties decreased (harder and less chewy structures).
			Meat products	
			Roast beef	The 6% inulin concentration provided the best sensory characteristics.
			Bologne sausages	
				Maintained the sensory properties and acceptability.
				Nutritional value in PUFA improved, the fat content decreased, and the fiber content increased.

Reduced Salt Content

The reduction of the salt content unfavorably affects some quality parameters of meat products such as sausages, bacon, cooked ham, and salami. In order to minimize the negative effects related to its texture and the interaction of water and fat, binding agents such as phosphates, lactates, chlorides, alginate, and transglutaminase have been used, thus preparing various meat products using any of these alternatives. The addition of phosphates in low-salt meat products improves the sensory and physico-chemical properties, since it increases the water and fat retention capacity, and reduces the salt content by up to 50%. Gel-forming agents, such as calcium alginate or the enzyme transglutaminase, improves the binding properties and texture. The combination of different sodium, potassium, and magnesium salts have been found to produce meat products with acceptable sensory quality characteristics.

Fat Content Modification

The reduction of fat content is generally based on the use of leaner meat or the addition of water and resistant starches, non-starch polysaccharides, gums, or proteins. For the development of low fat products, initial composition, desired final composition (percentage and type of fat) and the type of processing (cooking, curing, smoking, etc.) should be taken into account, since these factors affect the different quality attributes of the final product. The main disadvantages of reducing the fat content in meat products are the loss of juiciness and obtaining a hard and rubbery texture. The solution is the use of different combinations of vegetable fats, proteins, and carbohydrates as fat substitutes, which mimic the mouth feel and texture of the fat.

Reducing fat content in meat products does not reduce cholesterol content, and it has even been suggested that when fat is reduced and lean meat is increased, the cholesterol content of the meat product may increase. Development of meat products with less cholesterol is based on replacing fat and lean meat with vegetable products that do not contain cholesterol, such as vegetable oils and plant proteins.

Meat products with a more suitable composition can be obtained by modifying the fatty acid profiles by using fats of vegetable and marine origin as partial substitutes for meat fats. In general, vegetable oils are rich in MUFA and PUFA and contain bioactive compounds. The fatty acid composition of the reformulated meat product will be affected by the type of oil used. Various meat products have been elaborated using vegetable oils from olive, sunflower, cottonseed, corn, soybean, flaxseed, rapeseed, peanut, etc., and fish oils. Although the replacement of animal fat by vegetable oils improves the lipid profile of the products, the percentage of fat that can be replaced without negative effects needs to be investigated.

Researchers studied the effect of using different proportions of olive and linseed oils for the total or partial substitution of animal fat on beef patties. The best sensory results were obtained in beef patties when 50% of animal fat was replaced by 50% of a mixture of oils (25% olive oil and 75% linseed oil), which in turn gave rise to products with high content of n-6 and n-3 fatty acids of nutritional interest. Likewise, consumers found no differences in the sensory parameters of these patties, with improved lipid profiles, with respect to the conventional patties.

Nitrite Content Reduction

In order to inhibit the formation of N-nitrosamines derived from nitrates added to meat products, sodium ascorbate, and erythorbate have been tested, but their effectiveness is limited due to the low solubility in adipose tissue. Studies have been conducted on the addition of fat-soluble derivatives of ascorbic acid, such as L-ascorbyl palmitate and long-chain acetals of ascorbic acid, the combination of α-tocopherol and ascorbate, and the use of lactic acid to inhibit the formation of N-nitrosamines.

Until now, a single compound to replace nitrites has been impossible to find, due to the multi-functional role they perform in meat products. Therefore, the solution is to combine several compounds that affect the color, flavor, antioxidant, and antimicrobial activity. Dyes such as erythrosine or the natural coloring pigments formed during external curing ("mononitrosil ferrohemochrome") can be used as alternative methods for maintaining the color of nitrite treated meat. The taste imparted as a result of the added nitrites is due to its antioxidant activity, which is why different antioxidants and chelating chemical agents can be used for its replacement. To replace the antimicrobial effect of nitrites, numerous compounds such as sorbic acid, potassium acid, sodium hypophosphite, fumaric acid esters, parabens, lactic acid producing bacteria, etc. can be used.

Incorporation of Protein and Dietary Fiber

Plant proteins are used in meat products to reduce costs, and to improve nutritional benefits. Soybean and sunflower proteins, wheat and corn derivatives, cottonseed, and oatmeal flours have been used as fat substitutes in different meat products such as minced meats, hamburgers and sausages. The functions of plant proteins in meat products are that they act as binders, improve the binding of water and fat and improve the water retention capacity. Soybean protein has been used as a functional ingredient in different meat products like cooked minced meat and sausages.

Dietary fiber is incorporated into meat products due its health benefits and due to its ability to improve water and fat retention, increase emulsion stability, increase oxidative stability and modify texture. Prebiotics such as inulin, a soluble dietary fiber, and products rich in dietary fiber have been used in the formulation of fresh, cooked, fermented, and crude-cured meat products. The fiber-rich products that have been added come from many different sources, such as cereals, fruits, dried vegetables, roots, and tubers. Adding dietary fiber improves nutritional properties by decreasing fat content, increasing fiber content, and maintaining sensory features.

Enzymes

Enzyme applications include tenderizing meat, restructuring low-value pieces and trimmings of fresh meat for higher-quality products, and improving flavor and aroma. Table summarizes different examples in which they have been used and the effects on the sensory characteristics of meat products.

Table: Effects of some enzymatic treatments on the sensory characteristics of meat products.

Objective	Treatment	Meat Product	Effects
Tenderization	Addition of papain, bromelain, ficin	Beef meat	Increase of tenderness.
		Mortadella	Without changes in organoleptic properties.
		Turkey, hen and rooster thighs	
		Beef cubes	
	Blade tenderization, bromelain or salt/phosphate injection	Muscles from beef rounds	Injection with a salt and phosphate solution resulted in the lowest
			Warner-Bratzler shear force (WBSF) values.
			WBSF values for blade tenderization and enzymatic tenderization were comparable.
Restructuring	Addition of transglutaminase (TGase)	Restructured cooked ham	No effect on color.
		Pork gels	Formation of network structures, improving the textural properties: increase of springiness, firmness, decrease in adhesiveness.
		Low-salt chicken dumplings	
		Chicken sausages	Increased juiciness, tenderness and overall acceptability.
		Doner kebab of chicken	
		Sausages and ham.	Increase of firmness of meat gels.
		Restructured pork	
	Bisulfite, soybean protein and TGase	Pork sticks	Improvement of tensile strength and cooking performance.
	Sea spaghetti seaweed (3% dry matter) combined with NaCl reduction and a (TGase/ caseinate) system	Restructured poultry steaks	Increase in Kramer shear force.
			Products were acceptable.
Production of flavor and aromas	Addition of palatase M and protease P	Spanish dry fermented sausage (Pamplona chorizo)	Without changes in the sensory quality except a slight softening.
	Intracellular cell free extract (L lactis NCDO 763,) and α-ketoglutarate	Dry fermented sausages	Improvement of odor and flavor when L. lactis and α-ketoglutarate were combined.

Enzymes used for Meat Tenderization

Natural proteolytic enzymes that improve meat tenderness can be from plant, bacterial, or fungal origin. The most widely used plant proteolytic enzymes to improve meat tenderness are papain, bromelain, and ficin. The injection of papain into beef softens the meat, increasing tenderness, maintaining color, and organoleptic characteristics. Injection of beef fillets with a bromelain solution increases tenderness tenderizes meat. Treatment of mortadella with ficin softened the meat without modifying its organoleptic properties of the mortadella.

Enzymes used for Meat Restructuring

Transglutaminase (TGase) improves meat texture characteristics, binding, and performance parameters. TGase can be used in meat emulsions to increase the binding of the solubilized proteins forming a stronger network, increasing the stability of the emulsion. Moreover, TGase can bind meat of different shapes and sizes to obtain a uniform restructured meat product such as restructured cooked ham, low-salt chicken dumplings, chicken doner kebab, and restructured pork. The production of restructured meat products with TGase is usually combined with the addition of proteins such as sodium caseinate, bisulfite-treated soybeans.

Enzymes used to Produce Flavor and Aromas in Meat

The main enzymatic reactions that affect the flavor and aroma of meat products are proteolysis and lipolysis. These reactions can be carried out by endogenous proteases and lipases, enzymes of microbial origin naturally present in the product, or enzymes added during the manufacturing process. The use of enzymes to improve flavor and aroma has been used mainly in cured meat products. During the maturation of cured meat products, proteolytic enzymes break down proteins and produce nitrogenous compounds and precursors of volatile compounds, which contribute to the development of flavor and aroma. Lipases hydrolyze triacylglycerides into monoacylglycerides, diacylglycerides, and free fatty acids. Free fatty acids are oxidized to volatile aromatic compounds which contribute to the aroma of the final meat product.

The addition of proteases and lipases in the sausage and chorizo accelerates the proteolysis and lipolysis processes. However, researchers observed no improvement in flavor and aroma, and excessive softening occurred in the chorizo and in some sausages. The addition of an extract of Lactococcus lactis and α-ketoglutarate in the salami produced an increase in the content of volatile compounds, and improved the sensory properties of the salami, increasing the flavor and aroma of the salami.

Preservation of Physical Methods

Thermal Processing

Pasteurization or sterilization of meat at high temperatures to attain a better shelf life, improved palatability, and enhanced flavor is known as thermal processing. The time temperature combinations of thermal processing of meat is generally decided based on the required log reduction of the specific target microorganism, expected shelf life, and the physicochemical properties of meat. The main target microorganism in thermally processed RTE food generally is Clostridium botulinum. The target organism in processed meat is Listeria monocytogenes since it grows at refrigerated conditions. Cooking, sous vide cooking, canning, retort pouch processing, pasteurization is all different kinds of preservation techniques, which use high temperature to process and preserve meat.

Meat texture varies depending on the internal temperature applied during the processing and on the intrinsic characteristics of meat. The tenderness of meat increases at higher temperatures due to the denaturation of protein, solubilization of collagen, and formation of gelatin. Thermal processing inactivates endogenous proteolytic enzymes and prevents development of off-flavors due to proteolysis. Heat sensitive vitamins are lost during prolonged heating at higher temperatures, but a comparable increase in the shelf life, flavor, and palatability is associated with heat treated foods. Oxidation of sulfhydryl group to disulfide group imparts the cooked flavor to meat. The color of meat changes when meat is subjected to high temperatures since myoglobin gets oxidized, which increases redness and reduces lightness of meat, relating it to the doneness of meat. The consumer preference of the meat varies subjectively but the change in color is accepted as a preferred aspect of cooked meat.

Thermal processing reduces the oxidative stability of meat, which is detrimental to both nutritional and sensory quality of meat, but incorporation of antioxidants into meat has been proven to be a solution for the same. Heat treatment has been found to concentrate the micronutrients like zinc, magnesium, iron, phosphorous in meat; however, some amount of these micronutrients are lost to thermal leaching.

A lot of studies have been carried out indicating the possibility of a modification of intramuscular fat in meat leading to an increase in the proportion of unsaturated fatty acids and therefore the nutritional value of meat. RTE meat products tends to show acceptable sensory characteristics with a slightly reducing trend during storage; but these thermally processed foods in hermetic containers offers a shelf life of over 12 months with slight changes in sensory properties of meat. The sensory attributes like appearance, flavor, texture, juiciness, and overall acceptability of thermally processed traditional Indian meat curry, Rogan josh showed a significant declining trend during its shelf life of 12 months. Rogan josh is a meat product elaborate with thick gravy cooked with big chunks of meat added with spices and condiments, in the aforementioned study beef meat was used. The declining trend was due to the protein degradation and oxidation of the product. Similar results were observed by researchers in RTE retort pouch processed Indian delicacies like Chettinad chicken.

Packaging

Packaging options for meat and meat products are air permeable packaging, vacuum packaging, modified atmosphere packaging, active packaging, smart packaging, and edible coatings. Table summarizes different examples of packaging used in meat and meat products and their effects on sensory characteristics.

Table: Effects of different types of packaging on the sensory characteristics of meat products.

Treatment	Meat Product	Effects
Air-permeable packaging, vacuum packaging and modified atmosphere packaging	High pH and normal pH beef Lamb slices Beef steaks Beef fillets Lamb steaks Beef and pork steaks Bison tenderloin steaks	Vacuum packaging inhibits lipid oxidation, thus preventing unpleasant odors and flavors. Packaged in a modified atmosphere with semi-permeable internal vacuum film leads to an attractive bright red color. Packaging in CO_2 improves color and the stability of meat color compared to vacuum package.
Active packaging	Pork patties Beef steaks Beef	Active packaging does not affect the tenderness of the meat. Desirable bright red color. Incorporation of essential oils in the active packaging leads to unpleasant flavors and aromas.
Edible films and coatings	Beef Pork meat Minced beef Ready to eat (RTE) meat products Pork meat hamburgers Ground-beef patties	The stability of the red color is improved. Lipid oxidation is inhibited, thus preventing unpleasant odor and flavor.
Combination	Pre-cooked convenience-style foods: battered sausages, bacon slices, and meat and potato pies	Optical oxygen sensors in combined vacuum and modified atmosphere packaged and the use of ethanol emitters: ethanol flavor and aroma were not perceived by panelists in two of the three products assessed.

Vacuum Packing and Modified Atmosphere Packaging

Traditional packaging options for meat and meat products are air-permeable packag-ing, vacuum packaging, and modified atmosphere packaging. Modified atmosphere packaging improves meat color stability compared to vacuum storage, since the main-tenance of the bright red color of the meat is possible; however, increased lipid oxida-tion may occur due to the oxygen content incorporated in the atmosphere. Although vacuum packaging inhibits lipid oxidation, which prevents the development of unpleasant smells and taste, the color of the meat becomes purple, which is not the bright red color that consumers associate with fresh meat. Thus, second skin vacuum packaging is considered better than conventional vacuum packaging because it is more visually appealing and has less weight loss. Therefore, the combination of vacuum packaging and modified atmosphere methods can be an efficient way to reduce the negative quality changes that occur when using these systems separately.

Active Packaging

Active antioxidant packaging controls the oxygen levels of the packages. There are two active antioxidant packaging systems: separate antioxidant devices and packaging materials with built-in antioxidants. Stand-alone antioxidant devices are sachets, pads, or labels that contain oxygen scavengers. The active agent is incorporated into the walls of the packaging material, so that undesirable compounds are absorbed from the headspace or the antioxidant compounds are released into food. An advantage of incorporating antioxidants in packaging materials, compared to direct addition to food, is controlled release of the active compound.

Antimicrobial agents that have been used in active antimicrobial packaging include plant and spice extracts, essential oils, peptides, organic acids, antibiotics, bacteriocins, and silver ions. Essential oils have been widely studied as natural antimicrobial agents for meat and meat products, but they produce intense aromas and flavors that affect the sensory quality of meat. Therefore, novel technologies such as the encapsulation of essential oils in nano-emulsions, the incorporation of essential oils in nanorcils and the use of essential oils combined with other antimicrobial methods or agents have been developed. For instance, researchers designed an antimicrobial active packaging system for RTE meat products, consisting of a film of chitosan with thyme essential oil, which improved the color of the meat and prevented the appearance of unpleasant odors. Likewise, the use of chitosan coatings did not influence the sensory characteristics of chicken meatballs and chicken skewers or fresh pork sausages.

Intelligent Packaging

The most widely used smart devices in packaging are barcodes, radio frequency identification labels, time–temperature indicators, gas indicators, freshness indicators, and pathogen indicators. Oxygen indicators are the most widely used gas indicator in meat products, since oxygen can cause oxidative rancidity, color changes and development of pathogenic and altering microorganisms. The freshness indicators are based on the detection of freshness indicator metabolites, whose presence causes a color change.

This type of smart packaging does not prevent the degradation of the products, it only helps to maintain the physical and sensory characteristics, and inhibits degradation resulting from oxidation reactions or product contamination. Therefore, the combination of smart packaging with other technologies is necessary. Researchers studied the combination of smart and active packaging technologies and stated that this combination can be used to assist in the modification of conventional packaging systems in order to enhance product quality and safety. However, further studies would be necessary to assess the impact of the combination of smart packaging technology with other packaging technologies on the organoleptic quality of meat and meat products.

Edible Films and Coatings

The films are made with biopolymers that are based on hydrocolloids such as polysaccharides, proteins of animal origin and proteins of plant origin. Films formed are impervious to moisture and gases but have worse mechanical and functional properties than plastic films. Chitosan is a biopolymer with antioxidant and antimicrobial properties and can be used as a matrix to develop edible films. Edible films and coatings for meat packaging can be combined with the incorporation of active components with antioxidant and antimicrobial properties. The active components that can be incorporated in edible films are natural extracts, essential oils, natural polymers, protein hydrolysates, enzymes, and nanocomponents.

High Pressure Processing

HPP is also applied for preservation purposes here we focus on the conservation approach. Response of vegetative pathogenic and spoilage microorganisms to HPP depends on process parameters such as pressure, temperature, processing time, and on product parameters such as pH, water activity, salt content, and the presence of other antimicrobials. Inactivation of more than four log units of common vegetative pathogenic and spoilage microorganisms can be attained by HPP at 400–600 MPa with short processing times of 3–7 min at room temperature.

Ham samples treated by HPP showed increased hardness and syneresis during storage. Ham samples HPP treated at 600 MPa for 5 min showed 2 log and 3 log reductions of L. monocytogenes on the surface and interior respectively. Treatment at 600 MPa for 5 min in completely dry-cured ham reached the food safety objective for L. monocytogenes, without significantly affecting the physicochemical characteristics of dry-cured ham.

Preservation of Chemical and Biochemical Methods

Food Additives

The main additives used in the production of meat derivatives are antioxidants, binders, antimicrobials, curing agents, and curing accelerators. Synthetic antioxidants approved for use in meat products are butylhydroxyanisole (BHA), butylhydroxytoluene

(BHT), propyl gallate, terbutylhydroquinone (TBHQ), and tocopherols. These antioxidants delay or inhibit the oxidation of meat and meat products, and therefore avoid the appearance of unpleasant odors and flavors.

Binder additives are added to the meat to maintain a uniform dispersion of fat throughout the product and to prevent water loss during the different stages of processing, heating, storage, and cooling. The binder additives used in meat and meat products are phosphates, starches, xanthan gum, guar gum, sodium alginate, carrageenan, carboxymethylcellulose, etc. It is worth highlighting the functions of phosphates in meat products, which are to increase the water retention capacity and increase the stabilization of the emulsion. Phosphates also have other functions such as stabilizing color, inhibiting lipid oxidation, and promoting protein dispersion.

The synthetic additives used as antimicrobials in meat products are organic acids such as acetic, lactic, propionic, sorbic, benzoic, and citric acids, and sulfites. Sulfites have antimicrobial activity against decomposing microorganisms; however, sulfites cause health problems such as allergic reactions in sensitive people. Organic acids have activity against a wide variety of pathogenic and disrupting microorganisms. Sorbic acid is used in meat products for its inhibitory activity against yeasts and molds. However, it does not affect lactic acid bacteria, which makes it useful as a preservative in fermented meat products.

Nitrates and nitrites are the most widely used curing agent in meat products. Nitrites provide the red color and flavor of cured meat, and have antioxidant and antimicrobial properties. The reduction of nitrites and nitrates to nitric oxide is important for the color of cured meat. Nitrosyl myoglobin, which is the dominant pigment in cured meat products, is formed from the interaction of nitric oxide with the heme group of myoglobin. Curing accelerators such as sodium ascorbate, sodium erythorbate, ascorbic acid, or erythorbic acid are added to meat to speed up the curing process, as they reduce nitrites to nitric oxide. Nitrites react with amines and amino acids leading to the formation of N-nitrosamines, which are chemical agents with potentially carcinogenic, mutagenic and teratogenic activities. For this reason, alternatives are being sought to reduce or eliminate the addition of nitrites in meat products and reduce health risks.

Natural Antioxidant Ingredients

Natural antioxidant ingredients can be used as food additives in meat and meat products for their technological properties. Table summarizes the effect of the addition of natural antioxidants in the formulation of meat and meat products on their sensory characteristics. The antimicrobial and antioxidant activities of some plant extracts and/or their essential oils are mainly due to the presence of some major bioactive compounds, including phenolic acids, terpenes, aldehydes, and flavonoids. It should be highlighted that natural antioxidants can be incorporated into packaging systems.

Table: Effects of the addition of some natural ingredients on the sensory characteristics of meat products.

Natural Ingredients		Meat Product	Effects
Essential oils: thyme, oregano, pimento, clove, citron, lemon verbena, lemon, balm, cypress leaf. Raw chicken. Meatballs. Beef patties.		Lamb meat.	Extracts like clove or cinnamon increase of L^*, a^* and b^* values during storage.
		Concentrations of essential oils of oregano and thyme greater than 1% led to strong odor and unpleasant taste.	
		Clove extract increased the a^* values.	
Plant extracts: grape seed, green tea, pomegranate peel/rind, acerola, pine bark, bearberry, cinnamon bark, rosemary, garlic, oregano, sansho, ginger, sage.	Grape seed extract (GSE) and wine pomace.	Low sulfite beef patties.	GSE showed less color changes during the storage.
		Beef enriched with n-3 and CLA.	GSE can darken a meat product.
		Pork meatballs.	No modification of the sensory attributes except for the color.
		Chicken nuggets.	
	Green tea extract	Low sulphite beef patties.	No effects on odor, taste and texture.
		Pork meatballs.	Degradation of red color is delayed.
			No modification of the sensory attributes except for the color.
	Rosemary extract	Turkey meat.	No effects on sensory characteristics.
		Pork burgers.	
	Oregano extract	Sheep burgers.	No effects on sensory characteristics.
	Garlic	Low-salt sausages.	No effects on aroma, flavor and overall appearance.
	Other fruit extracts: blueberries, raspberry pomace, pomegranate peel and pomegranate juice.	Pork burgers and cooked pork ham.	Sensory quality was not negatively altered.
		Beef burgers.	
		Chicken patties.	
	Citrus fiber	Bologne sausage.	No effect on color or texture properties.
			When citrus fiber is combined with rosemary essential oil, the sensory parameters improved.
Spices Indian sheep meat product.		Rabbit burgers.	Turmeric powder leads to higher yellow values.
		Cumin and cardamom led to high overall accept.	

Essential Oils and Spices

In some cases, the addition of essential oils or spices can have negative effects on meat. For example, the addition of essential oils of oregano and thyme in lamb meat in concentrations greater than 1%, produces a strong odor and unpleasant taste in the product. Adding different extracts like clove or cinnamon to raw chicken can increase L^*, a^*, and b^* values during storage. Cinnamon also increases the redness of meatballs, although it does not affect the other sensory characteristics. The addition of turmeric powder to rabbit patties changed the color of the meat, due to the yellow color of turmeric. Therefore, studies of the concentrations of the added oils and spices, to avoid these negative effects on the sensory characteristics, are required. Adding clove extract to cooked and refrigerated beef patties increased the patties' red color and maintained the sensory characteristics for up to 10 days of storage. Addition of cumin, clove, and cardamom in rista, a traditional Indian sheep meat product cooked with spices, improved the shelf life of the meat product to 25 days with a high overall acceptability score.

Plant Extracts

The antioxidant activity of rosemary extract has been evaluated in pork burgers, and it was observed that it did not develop any adverse effects on sensory characteristics or general acceptability of the product. The addition of rosemary (0.25% v/w) did not negatively influence the taste of turkey meat. The incorporation of oregano extract in sheep burgers did not affect the sensory properties of the burgers. No comparable difference in aroma, flavor, and overall acceptability was observed in low-salt sausages when garlic derivatives were added to it.

The addition of green tea extracts to low sulfite beef patties delayed the appearance of rancid flavors, decreased the loss of red color, and did not modify the odor, taste, and texture of the patties. The addition of 250 mg/kg of grape seed extract did not affect sensory characteristics or instrumental color in beef enriched with n-3 and CLA. However, adding grape seed extracts and green tea could darken pork meatballs.

Sensory quality was not negatively altered when blueberries were added to pork burgers and cooked pork ham, or when raspberry pomace extracts were added to beef burgers, or extract of pomegranate peel and pomegranate juice in cooked chicken patties. The wine pomace, also called grape pomace, a by-product from winery rich in antioxidants is used in the preparation of meat products. For instance, red wine pomace may be an alternative to sulfites as a meat additive for protection of beef patties against protein oxidation. However, the presence of anthocyanins in red grapes has the disadvantage of darkening the product, which could modulate consumer opinion, what must be studied for every type of product. For example, the excess color of seasoned nuggets did not adversely affect the evaluation of other sensory characteristics such as juiciness, crispness, oiliness, saltiness, and chicken flavor.

The by-products of citrus fruit juice processing can be considered as potential ingredients in meat products because of their ability to reduce residual nitrite levels, thus avoiding the possible formation of nitrosamines and nitro samides. The addition of citrus fiber washing water did not affect the color or texture properties of Bologna sausage, and its combination with rosemary essential oil led to the best sensory quality.

6

Spices and Condiments

Dried seeds, roots or vegetables added to a food item in order to add flavor to it are termed as spices. Condiments constitute sauces, spices or any other preparation to complement a dish or enhance its flavor. This chapter discusses in detail the technologies related to spices and condiments as well as their chemical composition.

Spices can be defined as the dry part of a plant, such as roots, leaves, bark, fruits, flowers and seeds which are used as food adjuncts to add aroma and flavour to food products while condiments are also plant parts which are used as food adjuncts to add taste only. The flavour and taste attributes come into the foods due to the presence of essential oils in spices and condiments. These may be used as whole, ground, paste or liquid form for flavouring and seasoning foods.

Characteristics of Spice

- They are usually dried and ground to be mixed with other ingredients.
- They are having carminative properties, thus stimulate digestion.
- They add flavour and aesthetic, aromatic and therapeutic treatments to food, drink and other items.
- Dry varieties increase the shelf life of food.
- They are having antibiotic properties and prevent bacterial growth.
- They preserve food like meat for year or more without refrigeration.
- Some varieties improve texture while some introduce palatable colour or odour.
- They are of little nutritive value.

Classification of Spices

There are about 35 spices and condiments which can be broadly classified into 6 groups, based upon the parts of the plants from which they are obtained:

- Rhizomes and root spices: Ginger, Turmeric, and Garlic.
- Seed spices: Nutmeg, Coriander, Fennel, Cumin, Fenugreek, Dill, Aniseed, Celery and Bishop-weed.

- Bark spices: Cinnamon.
- Fruit spices: Cardamom, Black Pepper, Vanilla, All spice, Cassia, Tamarind, etc.
- Leaf spices: Bay leaf, Curry leaf, etc.
- Flower spices: Clove, Saffron, Asafoetida, etc.

Generally, spices are the primary ingredients used in dishes during their preparation, while condiments are used at the table to enhance the taste of the dish. In other words, spices are the main ingredients used during the preparation of a dish to impart flavour, aroma and piquancy to the foods while condiments are the secondary ingredients that are used after the dish is prepared and adds only taste to the dish e.g., ketchup.

Spices and condiments are major commercial crop and earn a major part of foreign exchange annually. They have been the backbone of agricultural industry. The importance of spices and condiment in dietary, medicinal and other uses, and their commercial importance are immense. Spices undergo following processing before consumption:

- Spice cleaning
- Spice reconditioning
- Spice grinding
- Spice packaging

Types of Spices

Turmeric

It is a rhizomatous herb plant of the ginger family. It is one of the key ingredients for many Indian, Persian and Thai dishes such as in curry and many more.

Cardamom

Cardamom refers to several plants of the similar genera Elettaria and Amomum in the ginger family. Cardamom is native to India. Cardamom was used as a medicine as well as a food. It became common in Europe in the 17th century when it was brought by sea from India. Cardamom is used as flavourings in both food and drink.

Chilli

Chillies are native to South and Central America and they have been cultivated for thousands of years. However in the 16th century they were introduced into South Asia and Southern Europe by the Spanish and Portuguese. Chilli powder became very popular in Britain in the 19th century.

Cinnamon

Cinnamon is a spice obtained from the inner bark of several trees from the genus Cinnamomum that is used in both sweet and savoury foods. Cinnamon bark is widely used as a spice. It is principally employed in cookery as a condiment and flavouring material. Cinnamon is native to Sri Lanka. Like many spices cinnamon was used as a medicine as well as a food. It was used to treat many complaints including colds and flu.

Cloves

Cloves are the aromatic dried flower buds of a tree in the family Myrtaceae. Cloves are native to the Maluku islands in Indonesia and used as a spice in cuisines all over the world. Cloves are harvested primarily in Indonesia, India, Madagascar, Zanzibar,

Pakistan, and Sri Lanka. They have a numbing effect on mouth tissues. Cloves are native to Indonesia but they were known to the Romans and to the Ancient Chinese. Cloves were also used as a medicine. Cloves are now grown in East Africa and in Brazil.

Coriander

Coriander also called cilantro or dhania, is an annual herb in the family Apiaceae. Coriander is native to southern Europe and North Africa to south-western Asia. All parts of the plant are edible, but the fresh leaves and the dried seeds are the parts most commonly used in cooking. Coriander is common in South Asian, Middle Eastern, Central Asian, Mediterranean, and Indian dishes. Coriander was grown by the Greeks and the Romans. It was also popular in Medieval Europe. For centuries coriander was used as a medicine and it was believed to be an aphrodisiac.

Cumin

Cumin is a flowering plant in the family Apiaceae, native from the east Mediterranean to India. Its seeds are used in the cuisines of many different cultures, in both whole and ground form. Cumin seeds are used as a spice for their distinctive flavour and aroma. It is globally popular and an essential flavouring in many cuisines. Cumin is native to the Mediterranean region and it was well known in Ancient Egypt as well as in Greece and Rome. Cumin was also a common spice in the middle Ages in Europe. It was also used as a medicine.

Garlic

Garlic is native to Central Asia. It was known in Ancient Egypt and also in India and China. Garlic was also grown by the Greeks and the Romans. For centuries garlic was used as a medicine as it is a strong antiseptic. Garlic was also believed to be an aphrodisiac. It originates in Yunnan province in Southern China.

Ginger

Ginger or ginger root is the rhizome of the plant Zingiber officinale, consumed as a delicacy, medicine, or spice. It lends its name to its genus and family (Zingiberaceae). Young ginger rhizomes are juicy and fleshy with a very mild taste. Ginger is native to Southeast Asia. It was known in ancient China and India and from the Middle Ages was used in Europe. However for centuries ginger was used as a medicine far more than for flavouring food.

Mint

It is a popular herb in Europe in the middle ages. Like many herbs mint was used as a medicine as well as a food. It was well known to the ancient Greeks and Romans.

Mustard

It was well known in the ancient Middle East and it is mentioned in the bible. The Romans also grew mustard and they introduced it into the parts of Europe they conquered.

Nutmeg

Nutmeg and mace have similar sensory qualities, with nutmeg having a slightly sweeter and mace a more delicate flavour. Mace is often preferred in light dishes for the bright orange, saffron-like hue it imparts. Nutmeg is used for flavouring many dishes, usually in ground or grated form, and is best grated fresh in a nutmeg grater. Nutmeg comes from Indonesia. It is having many medicinal uses.

Soya Sauce

Soy sauce is a condiment produced by fermenting soybeans with Aspergillus oryzae mold, along with water and salt. Soy sauce is a traditional ingredient in East and Southeast Asian cuisines, where it is used in cooking and as a condiment. In more recent times, it is also used in Western cuisine and prepared foods.

Chemical Composition of Condiments

Vinegar

It is a liquid substance consisting mainly of acetic acid (CH_3CO_2H) and water. Acetic acid is produced mainly through the fermentation of ethanol by acetic acid bacteria. Generally vinegar contains extract, ash, volatile acid, non-volatile acid, reducing sugar, the concrete substances present are higher alcohols, 3-hydroxybutanone, dihydroxy acetone, tyrosol, acetaldehyde, formaldehyde, acetal, acetic acid (the content is 3 to 5 %), succinic acid, oxalic acid and sorbose.

Thick Sauce

In every 100g of thick sauce, there are:

Water	39g
Protein	20.9g
Fat	11.2g
Carbohydrate	2g
Ash	24.9g
Calcium	245mg
Phosphorous	174mg
Iron	16.1mg
Thiamine	0.05mg
Ovaflavin	0.78mg
Nicotinic acid	2.1mg

In every 100g of sweet sauce made from fermented flour, there are:

Water	47g
Protein	5.8g
Fat	1.2g
Carbohydrate	37g
Ash	6.3g
Calcium	32mg
Phosphorous	104mg
Iron	5.7mg

Hot Sauce

There are many recipes for hot sauces - the common ingredient being any kind of chilli pepper. A group of chemicals called capsaicinoids are responsible for the heat in chilli peppers. The peppers are infused in anything from vinegar, oil, water, beer and alcohol to fruits and vegetable pulp. Additional ingredients are often used, including those used to add extra heat, such as pure capsaicin extract and mustards.

Olive Oil

Olive oil is a fat obtained from the olive fruit. The oil is produced by grinding whole olives and extracting the oil by mechanical or chemical means.

- Saturated fats:
    - Palmitic acid: 7.5–20.0%
    - Stearic acid: 0.5–5.0%
    - Arachidic acid: <0.2
    - Behenic acid: <0.3%
    - Myristic acid: <0.05%
    - Lignoceric acid: <0.2
- Monounsaturated fats:
    - Oleic acid: 55.0–83.0%
    - Palmitoleic acid: 0.3–3.5%
    - Linoleic acid: 3.5–21.0 %
    - α-linolenic acid: <1.0%

Mustard

Mustard is often used at the table as a condiment on meat.

Table: Nutritional value per 100 g (3.5 oz).

Energy	276 kJ(kcal)
Carbohydrate	8g
Sugar	3g
Dietary fiber	3g
Fat	3g
Protein	4mg
Magnesium	49mg (14%)
Sodium	1120mg(75%)

Table Salt

It is composed mainly of sodium chloride. It is a refined salt containing about 97 to 99% sodium chloride (Nacl) and anti-caking agents like sodium alumina-silicate or magnesium carbonate to make it free-flowing. Some other anti-caking agents include tricalcium phosphate, calcium carbonate, calcium silicate, sodium alumina-silicate and calcium alumina-silicate. However additives may also be added. Iodine containing compounds are added to table salt like potassium iodide, sodium iodide, or sodium iodate. Small amount of dextrose is added to stabilize the iodine. Iodized salt is used to reduce the incidence of iodine deficiency in humans.

Wasabi

The volatile for pungency is allyl isothiocyanate produced by hydrolysis of natural rhizome thioglucosides and the reaction is catalyzed by myrosinase.

100g of wasabi root contains:

- Calories: 109 kcal
- Fat: 0.63 g
- Carbohydrates: 23 g
- Fibre: 7.8 g
- Protein: 4.8 g

Ketchup

Ketchup has health benefits which are offset by the salt and sugar content. Ketchup is a source of lycopene, an antioxidant which may help prevent some forms of cancer.

Table: Nutrient (per 100 g).

Energy	100 (kcal) 419kJ
water	68.33g
Protein	1.74g
Fats	0.49g
Carbohydrate	25.78g
Sodium	1110mg
Vitamin c	15.1mg
Lycopene	17.0mg

Tamarind

Contains plant acids (16–18%) composed mainly of d-tartaric acid, with minor amounts of lmalic acid. Other constituents include polyphenolics (catechin, epicatechin, and procyanidin), flavonoids (taxifolin, apigenin, luteolin, and naringenin), sugars (20–40%), pectin, protein (2.8%), fat, vitamins (e.g., B1 and C), minerals (Ca, K, P, etc.), and tartrate. It also contains a volatile fraction that consists of over 60 identified compounds, including limonene, terpinen-4-ol, neral, α-terpineol, geranial, and geraniol, which are responsible for its citrus taste; methyl salicylate, safrole, ionones (β- and γ-), cinnamaldehyde, and ethyl cinnamate, which contribute to warm spicy taste; piperitone; and several pyrazines and alkylthiazoles are also present.

Table: Proximate composition per 100g.

Constituent	Percentage
Moisture	15.00-30.00
Proteins	2.00-9.10
Fat/oil/lipid, crude	0.50-3.10
Carbohydrates	56.70-82.60
Fibre, crude	2.20-18.30
Tartaric acid	8.00-18.00
Reducing sugars	25.00-45.00
Total ash	2.10-3.30
Pectin	2.00-4.00
Cellulosic residue	19.40
Albuminoids	3.00-4.00
Total available carbohydrates	41.77
Alcohol insoluble sugars	22.70
Water insoluble sugars	20.50
Non-reducing sugars	16.52
Total sugars	41.20-58.7
Starch	5.70
Tannin	600.00 mg
Ascorbic acid	3.00-9.00 mg
Carotene equivalent	10.00-60.00 μg
Thiamine	0.18-0.22 mg
Riboflavin	0.07-0.09 mg
Niacin	0.60 mg

Chemical Composition of Spices

- Basil: Estragole is an isomer of anethole, the compound found in anise. Basil oil contains a large quantity of estragole, along with the compound linalool. It is also found in herbal medicinal products.
- Bay Leaves: 1,8-cineole is the major constituent, a compound also found in cardamoms. Its common name is eucalyptol, the major compound found in eucalyptus oil. Eucalyptol is an ingredient added to cigarettes and is also found in many brands of mouthwash.
- Caraway: Carvone, the major compound in caraway, actually has two mirror image isomers, one of which smells like caraway, the other of which smells like spearmint. Only S-carvone, the isomer smelling of caraway, is found in caraway seeds.

- Cardamom: 1,8-cineole is the major compound found in small cardamoms. The cardamom aroma is caused by the combination of this compound and another compound, alpha-terpinyl acetate.
- Cayenne: Capsaicin is the major compound present in chilli peppers and gives them their spiciness.
- Chives: Similarly to onions and garlic, sulfur containing organic compounds gives chives their flavour. One of the major contributors in chives is dipropyl disulfide.
- Cloves: Eugenol is one of the main compounds found in cloves. More minor constituents are also contributors to the characteristic odour, for example compounds such as methylamylketone and methylsalicylate. Eugenol is named after the scientific name for cloves, and along with its derivative compounds is used in perfumery & flavourings.
- Cinnamon: Cinnamaldehyde gives cinnamon its flavour and odour; 90% of the oil obtained from cinnamon bark is cinnamaldehyde. It is also used as flavouring in chewing gum, sweets, ice cream and beverages.
- Coriander Leaves: Once of the main components of cinnamon leaf oil extract is 2-decenoic acid. The composition also includes many different aldehydes, primarily those of 9-10 carbons in length.
- Coriander Seeds: The main compound in coriander seeds is linalool. It has two mirror image isomers, one of which is known as coriandrol. The other mirror image isomer is found in lavender and sweet basil.
- Cumin: Cuminaldehyde is the main contributor to cumin's warm aroma. Other constituents include a range of other aldehyde compounds.
- Dill: Carvone is also one of the main compounds found in dill. The spearmint isomer of carvone is used in the manufacture of chewing gum – spearmint chewing gum is produced by being soaked in carvone.
- Ginger: Zingiberene is the major organic compound in ginger. The pungency of ginger is caused by a range of compounds called gingerols.
- Lemongrass: Citral is a mix of two different isomeric aldehydes, neral and geranial. Citral is also used in perfumery for its citrus odour.
- Mace: Mace's chemical composition is similar to that of nutmeg, as they are both obtained from the same plant. The compound of the highest concentration in the essential oil of mace is terpinen-4-ol. It is also found in the essential oil of the tea tree.
- Marjoram: Sabinene hydrate is the main component of the extracted oil of

marjoram, and is responsible, along with other compounds of the terpene family, for the characteristic flavour of the herb.

- Mint: The major compound in mint leaves is menthol. This compound is also a popular flavouring for chewing gum and toothpaste, and is also used in menthol cigarettes.
- Nutmeg: Sabinene is one of the major constituents of the essential oil of nutmeg. Another compound found in nutmeg is myristicin, which can cause hallucinogenic effects in large amounts.
- Oregano: Carvacrol is responsible for the warm, pungent odour of oregano. It is also found in tequila, and the oil of thyme.
- Paprika: Capsanthin is one of the compounds that contribute towards the red-orange colour of paprika, as well as the compound capsorubin. These two compounds increase in concentration during the advanced stages of ripeness of peppers, increasing the intensity of their red colour.
- Parsley: 1,3,8-p-menthatriene is the major compound found in parsley leaves. Other compounds include myristicin & limonene.
- Pepper: Piperine is the major constituent of the oil that can be extracted from black pepper, and is the main compound that gives black pepper its pungency. An isomer of piperine, chavicine, also contributes.
- Rosemary: Alpha-pinene is found in many species of coniferous trees, notably the pine tree, as well as being one of the major compounds in rosemary. Along with other members of the monoterpene family of compounds, it is emitted in large amounts by vegetation.
- Saffron: Crocin is the chemical compound responsible for the colour of saffron. It is a deep red colour, and forms an orange solution when dissolved in water. It has been shown to be a potent anti-oxidant.
- Sage: Manool is one of the main chemical compounds found in sage. Others include eucalyptol, and thujone, a psychoactive compound.
- Star Anise: 80-90% of the essential oil extract of star anise is anethole. This can be isolated and used to flavour liqueurs such as sambuca, as well as liquorice sweets.
- Tarragon: Estragole is an isomer of anethole, and is found in tarragon as well as basil. Estragole is suspected to be carcinogenic and genotoxic, but only at levels 100-1000 times the expected human exposure to the compound.
- Thyme: Thymol is the compound that provides the distinct flavour of thyme. It is also used as one of many additives in cigarettes.

- Turmeric: Compounds called curcuminoids are responsible for the yellow colouration of turmeric. They are also natural antioxidants. As a food additive, curcumin's E number is E100.
- Vanilla: The aroma of vanilla is mainly due to the compound vanillin, which accounts for 74-96% of the flavour & aroma compounds. Over 100 other volatile compounds have been detected, including acids, phenolic compounds, alcohols and aldehydes.

Technology of Manufacturing Spice Oil and Oleoresins

Oleoresins are flavoured compounds obtained by the solvent extraction of the ground spices. They have aroma and possess and attributes which contribute to the taste like pungency. All spices contain essential oils in various proportions which can be extracted by steam distillation. Oleoresins are preferred because of their microbiological advantages and uniformly in flavour and pungency, easy to store and transport. They have several applications like preparations of beverages, soup powders, curry powders, confectioneries, noodles, canned meats, sauces.

Oleoresins have large domestic as well as export markets. They are consumed by broad spectrum of manufacturer like beverages, soup powders, curry powders, confectioneries, noodles, canned meats, sauces, poultry products and so on. Most of the end use industries are growing steadily and bound to increase with increase preference of quality products. The use of spice is rapidly replaced with oleoresins and exports of these products instead of raw spices results in considerable value addition.

Types

Ajowan seed oleoresin, black pepper oleoresin, capsicum (chilli) oleoresin, cardamom oleoresin, cassia bark oleoresin, celery seedoleoresin, cinnamon bark oleoresin, clove bud oleoresin, coriander seedoleoresin, cumin seed oleoresin, curcumin powder, date extract, fennel seed oleoresin, galangal oleoresin, ginger oleoresin, juniper berry oleoresin, mace oleoresin, nutmeg oleoresin, olibanum resinold parsleyseed oleoresin, turmeric oleoresin, white pepper oleoresin, zingiber oleoresin, paprica oleoresin.

Manufacturing Process

Various raw spices are cleaned and then ground to require meshing size. The extraction is undertaken with the help of proper solvent. The solvents that can be used are hexane, acetone, ethylene dichloride or alcohol. Extraction is done by percolation of the solvents at room temperature through a bed of ground spices packed in a SS percolator.

The dark viscous extract containing not less than 10% of total soluble solids are drawn off and distilled under reduce pressure to remove the excess of solvent. The essential oil is obtained by steam distillation.

Paprika Oleoresin

(Also known as paprika extract) is an oil soluble extract from the fruits of Capsicum Annum Linn or Capsicum Frutescent, and is primarily used as a colouring and/or flavouring in food products. It is composed of capsaicin, the main flavouring compound giving pungency in higher concentrations, and capsanthin and capsorubin, the main colouring compounds (among other carotenoids). Extraction is performed by percolation with a variety of solvents, primarily hexane, which are removed prior to use. Foods coloured with paprika oleoresin include cheese, orange juice, spice mixtures, sauces, sweets and emulsified processed meats. In poultry feed it is used to deepen the colour of egg yolks. In the United States, paprika oleoresin is listed as a colour additive "exempt from certification" and generally classified as a natural colour. In Europe, paprika oleoresin (extract), and the compounds capsanthin and capsorubin are designated by E160c.

Turmeric Oleoresin

It is extracted from curcuma longa L., a member of ginger family. Obtained by solvent extraction of the ground spice, contains colouring matter, volatile-oil, fatty oil and bitter principles. It is orange-red in colour and consists of an upper oily layer and a lower crystalline layer. Colouring components are curcumin and curcuminoids. It is Fat and alcohol soluble, cold water insoluble; commercially dissolve curcumin in polysorbate-80 or –60 to make it water dispersible. It has good heat stability and poor light stability. The colour hue change with pH greenish in acidic pH, orange yellow in neutral pH, more stable in acidic pH than in neutral or alkaline pH. Used in pickle, bakery, confectionery pudding, gelatin, gummy bears, yogurt, popcorn and finger foods.

Black Pepper Oleoresin

- Description: Oleoresin Black Pepper is the natural extract of dried tender berries of Piper Nigrum Linn of family Piperaceae. It is obtained by the solvent extraction of Black Pepper and the solvent traces are removed by distilling it in vacuum at controlled temperature.
- Physical Appearance: It is a yellowish brown viscous liquid with pungent slightly biting aroma of Black Pepper. Spice Equivalent: About 5-10 kgs of Oleoresin replaces 100 Kgs of dried black Pepper.
- Packing: 10, 20, and 40 Kg HDPE Carbouys5, 10, 20, 40 Stainless Steel containers Shelf life and Storage: 18 months. Product should be stored in full tight containers at controlled atmosphere protected from direct heat and light.

Rose Mary Oleoresin

Rosemary Oleoresin Extract (also known as ROE) is not a preservative, it is an antioxidant - these slow the rancidity of Oils. If using to extend the shelf life of fixed oils (such as Hemp, Avocado etc.), it must be added to the oils when they are fresh before oxidation has started. It is also most effective if thoroughly dispersed in the oil. Rosemary Oleoresin Extract is a thick liquid, which can be difficult to blend. It is suggested to remove a small portion of the base oil, mix thoroughly, and then re-introduce this into the bulk fixed oil. Rosemary Oleoresin will also add shelf life to Sugar/Salt Scrubs, Balms and Moisturisers (Creams and Lotions) and it is added during the cool down phase or when the Preservative is added to the formula. Usage 2 - 10 drops per 100g of product Usage Rate: 0.1% 0.5%.

Ginger Oleoresin

On steam distillation, dried, cracked and comminute ginger yields 1.0 to 3.0% of pale yellow, viscid oil. The oil possesses the aromatic odour but not the pungent flavour of the spice. Of course, the odour of the oil is quite lasting. Ginger Oleoresin is obtained by extraction of powdered dried ginger with suitable solvents like alcohol, acetone etc. Unlike volatile oil, it contains both the volatile oil and the non-volatile pungent principles for which ginger is so highly esteemed. Concentration of the acetone extract under vacuum and on complete removal of even traces of the solvent used yields the so called oleoresin of ginger.

Ginger oleoresin is manufactured on a commercial scale and is in great demand by the various food industries. The aroma of ginger is pleasant and spicy and the flavour penetrating, slightly biting due to antiseptic or pungent compounds present in it, which makes it indispensable in the manufacture of a number of food products like ginger bread, confectionary, ginger ale, curry powders, soft drinks, vegetable, meat and fish curries, ginger cocktail, sauces etc. Ginger preserves and ginger candy prepared from green or fresh ginger is quite favourite of many and find great demand. A number of alcoholic beverages are prepared from ginger in foreign countries, such as ginger brandy, ginger wine, ginger beer etc.

Cinnamon Bark Oleoresin

Cinnamon oleoresin is prepared by extracting cinnamon bark with organic solvent. Oleoresin yield varies from 10 to 12 per cent. The oleoresin is dispersed on sugar, salt and used for flavouring processed foods.

Usage of Oleoresins

- Processed meats: The use of spices, particularly pepper, in the manufacture of meat products, is traditional to impart flavour and keeping quality to the products. Typical seasoning mix for fresh sausages, for example, consists of pepper, capsicum, ginger, nutmeg, plus herbs. For dry sausages and pickled meats cardamom and coriander are also used. The move to use oleoresins has been accelerated by the increasing size of the manufacturing plants, where the use of spice extracts benefits production quality, as well as easy handling and cost savings. The above spices are used in the dispersed form of their oleoresin, with cardamom and coriander in the form of their oils.

- Fish and vegetables: Seasoning mixes for both fish and vegetables, and particularly pickled or brined products such as herrings, contain a wide range of spices and herbs. The use of oleoresins, particularly dispersed oleoresins on a soluble base, will provide a means of easier preparation, reduced handling and costs.
- Soups, sauces, chutenys, and dressings: The increasing demand for convenience products available in the form of a dry mix for ready reconstitution has caused a rapid move from conventional seasoning towards dispersed or encapsulated oleoresins and oils. Oleoresins of celery, pepper, capsicum, are used in conjunction with the oils of onion and garlic. Coriander and ginger extracts are used in barbeque sauces. The uses of the lesser – known but highly useful spices such as fenugreek broaden the new product opportunities.
- Cheeses and dairy products: The use of spices in cheeses is established in Germany, including "Quark". Spice extracts are unlikely to be uses in these products as the spices provide the flavouring plus visual impact. However, spice oils and oleoresins will have significance in processed cheeses and savoury spreads.
- Baked goods: The use of cardamom in baked goods in Scandinavia and Germany is traditional. The baking industry generally uses ginger, cinnamon, and nutmeg. The move from the spice to their oleoresin has been effectively taking place for many years for ease of handling and simplicity in manufacture. The use of spice extracts in cake fillings, biscuits, and snack products is also increasing steadily.
- Confectionery: The use of spices and spices extracts in the confectionery area is rather rare, but demonstration of the use of such material as cardamom oil and other extracts in toffees, chocolates, and others, has shown that they provide a very novel and pleasing confectionery ingredient new to the market.
- Snacks: The flavouring is an essential component of the appeal of snack products and unusual because the flavour is often applied on the surface, either by spray coating or dusting. For this purpose the seasoning mix has to be capable of being applied in spray form, or powder. Oleoresins of pepper, chilli, and celery, are widely used. Turmeric and chilli extracts are used to provide colour.
- Beverages: Spice oils are used for the preparation of soft beverages, as for example ginger oil in the preparation of ginger beer, etc. Some of the less well-known spice extracts can be used to produce very pleasing soft drink products as yet not widely known outside of local production in countries where they originate.
- Cosmetics: The use of spice oils in the preparation of creams, soaps, shampoos,

lacquers, lipsticks, etc., is well known. The growing preference for herbal, spicy, and spicy coniferous products like shampoos and hair tonics are noted, yet such extracts as those of cardamom and fenugreek are little heard of. The use of lesser known spice extracts can provide new product appeal.

- Perfumes: Perfumery uses a wide range of essential oils and oleoresins from sources far and wide, and yet some of the lesser – known oils and oleoresins are hardly used at all.
- Hygiene products: Products like toothpastes, mouthwashes etc., depend on essential oils to provide their pleasing flavour, making them not only acceptable, but pleasant to use. In cleansing materials, detergents etc., spice oils provide the aromatic appeal in otherwise uninteresting and sometimes offensive notes associated with some of the base products.
- Aerosols: The use of aerosols worldwide is increasing at a significant rate in products such as air fresheners, polishes, lacquers and many cleansing agents, as well as waxes etc., All of these are perfumed with essential oils to provide their pleasant and fresh aroma.
- Pharmaceuticals: Both oils and oleoresins are widely used in pharmaceutical products, to provide either pleasant taste or aroma to render the medicinal products, which would otherwise be difficult to accept, pleasing and easy to use. These include medications, skin creams, cold remedies, etc.

Chillies

The chilli (Capsicum annum) is native to Mexico but is now grown and used extensively in most parts of the world. It is the most popular spice and is used throughout the world to flavour and add interest to bland foods. The seeds of chilli have a long shelf life of 2-3 years which helped in their global spread. Chiiiles belong to the same plant family (Solanaceae) as potato, tomato and aubergine. There are at least 150 different types of chilli, varying in the degree of hotness. They are rich in vitamin C, stimulate the appetite and cool the body, especially in hot climates, by making the person sweat.

Production

The chilli plant is a small bush that grows up to about 0.6m tall. It has white flowers that produce fruits in a variety of sizes and shapes. Some chillies (the cayenne pepper) are like stumpy fingers, while others such as the bird's eye chilli are tiny. The hottest chilli is the habanero which looks like a mini sweet pepper. Chilli plants row at altitudes from sea level to 1800 metered in the tropics. Their pungency is influenced by several factors such as high night temperatures and drought or over-watering. Green chillies are immature fruits and red chillies have been allowed to ripen for a further four weeks. Ripened chillies can also be orange-yellow, purple, dark brown or black.

Harvesting

By definition processing does not involve harvesting. However, one cannot produce a good quality product from badly harvested materials. Correct harvesting techniques are one of the most important factors in the production of a high quality final product. For processing, chillies should not be picked until they are mature and start to turn red.

Cleaning

The crop should be cleaned before processing. The first stage is to remove dust and dirt using a winnowing basket. This can be made locally from bamboo, palm or other leaves. Someone used to this work can remove the dust, dirt and stones quickly and efficiently (e.g. they could clean 100kg of chillies in an eight hour day). Small machines are available for cleaning but they are rarely cost effective. After winnowing the crop needs to be washed in clean water. All that is needed are two or three 15 litre buckets. For larger quantities a small sink needs to be constructed. This can be made out of concrete. However, the water must be changed regularly to prevent recontamination by dirty water. Only potable water should be used. Take care not to over-wet the chillies or they will take longer to dry.

Drying

This is by far the most important stage of the process. If the chillies are not fully dried or if they take a long time to dry, they will be prone to mould growth and spoilage. The sale value of mouldy chilli can be less than 50% the normal value. In extreme cases the whole crop can be lost.

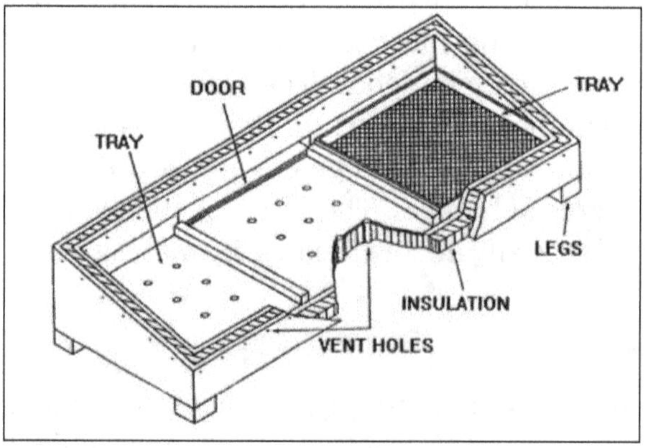

A cabinet solar dryer.

The choice of dryer will depend on the climate at the time of harvest and the intended end use of the chillies. For home use of the dried chillies, it is preferable to use the cheapest method available, which is sun drying. However, sun drying is really only practical in dry climates with plenty of sunshine. In humid climates drying will take too long, during which time the chillies have the potential to spoil. If a solar dryer is available, it is advantageous to use it as the drying process will be speeded up and the end result will be a higher quality dried chilli. Artificial dryers are only an option if there is a guaranteed market for the dried chillies. The Practical Action technical briefs on drying give a good overview of the principles and practicalities of drying and good advice on the choice of dryer.

Dying During the Dry Season

During the dry season, sun drying is usually adequate to dry the produce. The simplest and cheapest method is to lay the produce on mats in the sun. However, there are problems associated with this method. Dust and dirt are blown onto the crop and unexpected rain storms can re-wet the crop. To improve the cleanliness of the process, the drying chillies should be covered with a light gauze or muslin sheet which keeps away the insects and dust. To help maintain a goof red colour, the chillies should be dried in the shade, not in direct bright sunlight.

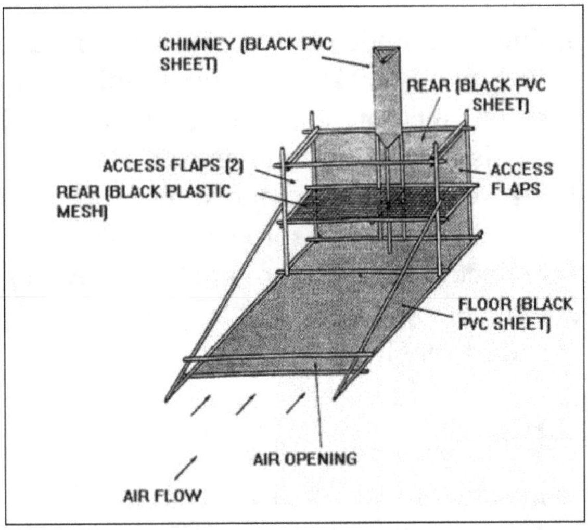

The 'Exell Solar Dryer'

Alternatively a solar dryer can be used. The simplest type of solar dryer is the cabinet solar dryer, which can be constructed out of locally available materials (e.g. bamboo, coir fibre or nylon weave). For larger units (over 30kg/day) an 'Exell Solar Dryer' could be used. However, the construction costs are greater and a full financial evaluation should be made to ensure that a higher income from better quality spices can justify the additional expense.

Drying During the Wet Season

During the wet season or times of high humidity, which often coincides with the harvest of the spices, a solar dryer or sun drying cannot be used effectively. An artificial dryer which uses a cheap energy source is necessary. This may be a wood or husk burning dryer or a combined wood burning and solar dryer. The technical brief on drying of foods contains more information on artificial dryers.

- Over drying: Care needs to be taken to prevent over drying of the chillies. A dryer operator will soon learn how to assess the moisture content of the chillies by hand. The final moisture content should be 10% wet basis.
- Grading: In some cases the dried chillies need to be graded, e.g. to gain a premium price for high quality packaged products. Chillies are graded by colour and size – the brighter the colour red the better. Grading is carried out by hand.

Grinding

Dried chillies can either be sold whole or ground into a powder. Grinding is one way of adding value to the product, but it must be done carefully to avoid problems and losses of material. A whole, intact product can be easily assessed for quality whereas a ground product is more difficult. Some consumers do not like to buy ground spices for fear that they have been adulterated. This fear can be overcome by producing a consistently high quality product and gaining the confidence of customers. Ground spices also lose their flavour and pungency much more quickly than whole spices. Therefore, grinding the chillies reduces their shelf life and potentially means that you may suffer from higher losses as you will have to discard any ground spice that does not sell. It is essential that the ground powder is well packaged in moisture proof bags to prevent it taking up moisture. You really should only grind spices if you have an assured market with a rapid turnover for the product. Basically there are two types of grinder - manual and mechanical. Whichever type you opt for, it must be placed in a separate and well-ventilated room because of the dust that it creates.

Manual Grinding Mills

There are many manual grinders that could be used to grind chilli. An experienced operator can grind about 20kg in an eight hour day. However, this is hard and boring work. A treadle or bicycle could easily be attached to the grinder which will make the work easier. With this system one person could grind about 30kg in one day.

Consumer research should be carried out to find out the fineness of grinding that the consumer wants. The grinding mills then need to be set so that they produce the desired ground product. For small-scale production, (up to 100kg/day) a series of these grinders is all that is needed. For larger scale production units, a mechanical grinder would be required.

Mechanical Grinding Mills

There are a range of mills - horizontal plate, vertical plate or hammer mills - that are suitable for grinding chillies. The choice of mill depends on what is available in your particular area and the price of the mill.

Packaging

- Packaging material: Packaging of these products, especially if they are ground requires polypropylene. Polythene cannot be used as the flavour components diffuse through it.
- Simple sealing: The bags can be sealed simply by folding the polypropylene over a hacksaw blade and drawing it slowly over the flame of a candle. However, this is extremely uncomfortable as the hacksaw blade heats up and burns the hands of the operator. It is however a very common technique and one that works adequately to seal the bags.
- Sealing machines: A sealing machine will considerably speed up this operation and produce a much tidier finish (which is very important). The cheapest sealing machines have no timing mechanism to show when the bag is sealed and they have a tendency to overheat. It is desirable to have a sealing machine with a timer. These machines come in many sizes. For most work an 8 inch (20cm) sealer is sufficient. Eye catching labels should be sealed above the product in a separate compartment and holed so the package can be hung-up in the shop.

Storage

A well designed and secure store is essential. The optimal conditions for a store are: low temperature, low humidity and free from pests. The store should be located in a shaded, dry place. To keep humidity as low as possible only fully dried products should be stored in it. The produce should be checked regularly and if it has absorbed too much moisture it should be dried again. To prevent pests entering, the roof should be completely sealed. Mosquito netting should be placed over the windows and doors should be close fitting.

Cardamom

Cardamom is a spice made from the seed pods of various plants in the ginger family. Cardamom pods are spindle-shaped and have a triangular cross-section. The pods contain a number of seeds, but the entire cardamom pod can be used whole or ground. The seeds are small and black, while the pods differ in color and size by species.

Green Cardamom

Green Cardamom is also referred to as “True Cardamom.” It’s the type of cardamom you’re most likely to encounter in the grocery store. It will come in the form of a green, paper-like pod. Inside the pod will be cardamom seeds. More than likely if your recipe calls for “cardamom” it’s referring to green cardamom. Of all the types of cardamom, green cardamom is the most well rounded, able to easily be incorporated in both sweet and savory dishes & drinks.

Bleached Cardamom

Bleached cardamom is actually green cardamom which has gone through a bleaching process. The idea behind this bleaching process is the tone down the flavor usually present in green cardamom, making it a more subtle flavor and smell. Some recipes might call for this type specifically to reduce the cardamom flavor, but in most cases, it's not preferable to green cardamom.

Black Cardamom

Black cardamom is not, technically, cardamom but it's a very commonly used close relative of green cardamom. Unlike green cardamom, black cardamom has an earthier and smokier flavor and smell, which makes it less flexible compared to its green counterpart. While green cardamom can often safely be used in sweet recipes, black cardamom cannot. That being said, black cardamom can be utilized to really add a unique depth of spice and flavor to any number of savory dishes which is unable to get through any other means.

In a tight pinch, you could possibly use green cardamom in place of black cardamom but you should never attempt to use black cardamom in place of green cardamom. You would be better off looking to substitutes for green cardamom, like cinnamon, nutmeg, or cloves, if you're unable to purchase cardamom itself.

Cardamom Seeds

Cardamom seeds come from green cardamom pods. There are usually 10 to 12 seeds per cardamom pod. While some recipes make use of whole cardamom seeds, many people purchase cardamom seeds in order to ground them personally, at home or in their restaurant. Like grinding whole peppercorns or coffee, the flavor and smell of cardamom is better preserved and expressed when you freshly ground your own cardamom.

Ground Cardamom

Ground cardamom comes from the seeds of green cardamom. It resembles other darker spices like cinnamon or nutmeg and is sometimes grabbed by mistake in the supermarket. Ground cardamom is the least flavorful and aromatic of all the types of cardamom as it has gone through an extensive process to arrive on the shelves ground and ready to consume.

A lot of the scent and flavor is found in the oils contained in the seeds, but as the cardamom is dried, ground and packaged, much of that oil is dried up and lost. Often ground cardamom will be used in recipes which are seeking to capture a more subtle cardamom flavor, like in a pie or baked good. There is a use for cardamom in just about every type of recipe: from aromatic blends and rich savory dishes made with black cardamom to the more subtle, complimentary flavors found in bleached and ground cardamom.

Drying

The drying is the most critical process in the production of dried herbs and spices. The aim of drying is to reduce the moisture content of the product from actively growing in the field to a level that prevents deterioration of the product and allows storage in a stable condition. Drying is a two stage process: firstly the transfer of heat to the moist product to vaporize the water in the product and secondly mass transfer of moisture from the interior to the product surface where it evaporates. The most important and immediate management concern is to ensure the harvested crop will not rot or become grossly invaded with yeasts, bacteria and mould (producing aflatoxins) or become contaminated by pests. This is the start of the preservation process, which for most spice crops requires drying that will enable the long-term crop storage and the opportunity for further processing. The drying phase of post-harvest management can include four preliminary stages - the selection of high quality produce from the field; cleaning the crop by washing and disinfection; preparing the crop for drying by peeling or slicing; pre-treating with anti-oxidants, blanching or sulfurizing. In some cases, washing prior to processing is desirable to remove field contaminants (dust, soil) using antimicrobial solutions to reduce the microbial populations to a low level prior to the drying process.

Cardamom capsule colour and the retention of greenness are important for highest quality. Capsules cured immediately after picking retained greener colour and the loss of greenness was more significant if the capsules were stored for more than 12 hours. Bagging of the capsules in jute bags, and cool storage, aids the retention of greenness. Cardamom capsules at picking contain 70-80% moisture and to enable storage the moisture has to be. Sun-drying is generally undesirable for cardamom because of bleaching and capsule splitting. The most widely-adopted drying system is a slow dry over 18-30 hours using a number of methods of artificial drying (electric, kiln, bin) with various methods of hot air flow. A temperature between 40-50 °C helps retain the green colour and an increase in temperature significantly increases the percentage of yellow capsules, split capsules, and heat injury.

There are four main types of drying. The most basic method of drying is to spread the crop on a surface exposed to the sun. In this case, the process is aided by a cover system that prevents wetting with rainfall. An improved method, speeding up the drying, is to use a fuel source (wood, oil/diesel, gas or electricity) to heat the drying room. Solar drying systems together with solar powered fans are also available. The drying process should dry the crop as quickly as possible, at temperature levels which do not drive off the volatile flavour compounds. The drying temperature regime will be specific to each crop as will be the final moisture percentage for storage. The traditional open sun drying that is widely used in developing countries has major inherent limitations when trying to preserve product quality. High crop loss and low product quality result from inadequate drying, long drying times, fungal spoilage, insect infestations, bird and rodent damage and contamination plus the effects of sunlight and the weather. Even in the most favourable climate it is often not possible to get the moisture content of the product low enough for safe storage. In the tropics the high relative humidity of the air prevents drying of harvested crop products during the wet season.

The objective of a dryer is to supply the product with more heat than is available under ambient conditions. A relatively small amount of heating can greatly enhance the moisture carrying capability of the air. For example, heating air from 20 °C at 59% relative humidity (RH) to a 35 °C at 25% RH increases the moisture holding capability three times. In a dryer the major requirement is the transfer of heat to the moist product by convection and conduction. The absorption of the heat by the product supplies the energy necessary for the vaporization of water from the product. The process that occurs at the surface of the product is simply the evaporation of the moisture. The moisture replenishment to the surface is by diffusion from the interior and this process depends on the nature of the product. Spices and essential oil crops derived from leaves or flowers are relatively thin and therefore relatively easy to dry due to their small diffusion thickness. Conversely in thick and fleshy materials such as roots, the drying process requires much more careful control of temperature, temperature ramp rate and airflow rate. If the temperature is too high, elevated too quickly or the airflow rate is too high when drying thick fleshy products, 'case hardening' may result, and only the outer surface will dry. This dry layer becomes impervious to subsequent moisture transfer.

In many rural locations, grid electricity and supply of other non-renewable sources of energy are too expensive, unavailable or unreliable and drying systems that use mechanical fans and electric heating are inappropriate. The high capital-and running-costs of fossil fuel-powered dryers presents an economic barrier to use by small-scale farmers. Solar-energy drying systems are put forward as technology for small rural farmers and enterprises in developing countries. Solar dryers can be classified into two generic groups, passive or natural air circulation solar dyers and active or forced convection solar-energy dryers.

Forced convective dryers employ motorised fans for circulation of the drying air. The electricity for the fan can come from a solar photoelectric panel and battery. Each group can also be sub-divided into three subgroups:

- Integral types (direct solar dryers where the crop is placed in a drying chamber with transparent walls, and the solar radiation falls directly on the crop, coupled with convection air flow from the heated surrounding air).
- Distributed type (indirect heating, where solar radiation heats a solar collector external to the drying chamber).
- A mixed type where there is both direct and indirect heating.

The integral type dryers are simple in both construction and operation and require little maintenance. However, they are likely to operate at lower efficiency due to their simplicity and there is less control of the drying operation. The distributed type are much more elaborate structures so require greater investment in materials and running costs, but have higher efficiency and as a result product quality is generally higher. The solar dryer has two significant disadvantages: a limited ability to process crops when the weather is poor; and drying can only occur during the daytime. This not only limits production and extends the drying time but also may have an adverse effect on production and product quality particularly fleshy crop products such as roots and stems that typically have drying times of several days. This has led to the development of hybrid systems with auxiliary heating systems such as burners using biomass, biogas or fossil fuels.

Alternatively, to achieve more efficient energy use, some active solar dryers are designed with thermal storage devices (mainly rocks or gravel) to extend the drying time during the night time and in periods of low sunshine. Desiccants (such as clay and rice husk) can be incorporated in the design to reduce the relative humidity of the drying air so as to improve the moisture carrying capacity. The use of desiccants is only possible in forced convection systems as they increase the resistance to airflow.

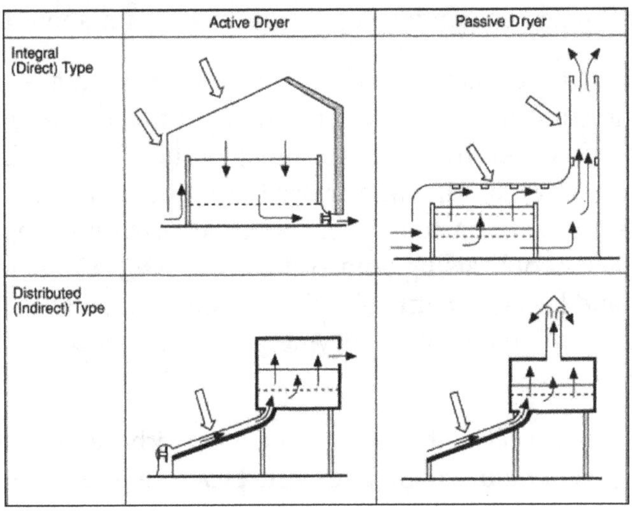

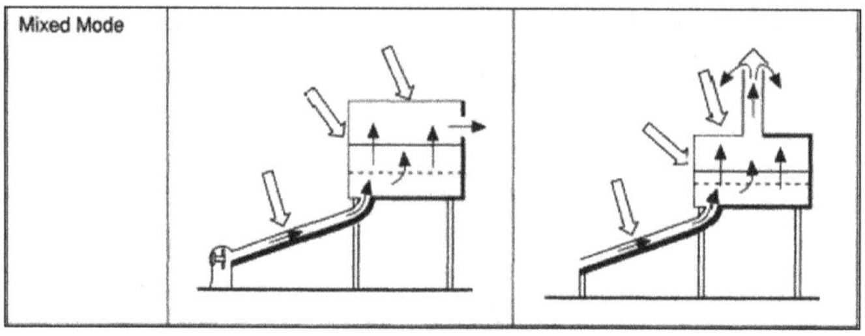

Typical Solar Dryer Designs.

Drying or Curing of Small Cardamom (Elettaria Cardamomum Maton)

Cardamom curing may be defined as the process in which moisture of freshly harvested capsules is reduced from 80 percent to 10-12 per cent through indirect heating. Drying is the most important unit operation that determines the colour of the end products. Apart from quality in terms of colour, flavour component such as 1-8 cineole, terpenyl acetate, linalool etc. are also important The traditional firewood based curing house consists of wood burning surface, flue pipes and drying racks in chamber. The furnace is made up of fine bricks or thick iron plates to facilitate charging with wood logs. Drying should be carefully controlled and should not be too rapid, temperature, humidity and aeration may be checked at frequent intervals during the course of drying.

The harvested capsules are spread in a single layer on trays. After keeping cardamom trays in the racks curing room is closed and heating is done by burning firewood in the furnace. The hot air passed through the pipes placed a few centimeters above the floor enhances the room temperature to 45 ñ 55 °C and this temperature status should be maintained for three to four hours initially. At this stage the capsules sweat and with the enhanced temperature, give off moisture. The ventilators are opened for sweeping out water vapour from the drying fruits. Exhaust fans can also be used for the speedy removal of moisture. After the complete removal of water vapour, ventilators are closed and the temperature inside the chamber should be maintained again at 45-55 °C for about 18 to 24hours.

At frequent intervals (three-four hours) ventilator exhaust fan are to be opened to remove the excess moisture otherwise water vapour settles down on the capsules and it leads to cooking effect thereby loses the colour of the capsules. The temperature is again raised to 60-65 °C for another one or two hour for completing the curing process. The temperature is raised to hasten the cleaning process removing debris like stalks attached the capsules. Avoid raising temperature above 65 °C inside the curing house to reduce splitting and loss of volatile oil. Black coloured blanket kept on the capsules spread under the sun will augment the drying process and also provides green colour to the cured produce.

Dried capsules have to be polished either manually or with the help of machines before marketing. Polishing is done by rubbing the dried capsule in hot state against the hard surface. Polishing machine is also available which can be operated either manually, or with electric motor. Motorized machines like Carpol can be used for polishing as well as grading of capsules having desired mesh. After grading, cardamom capsules can be stored over long duration. For efficient retention of green colour during storage, cardamom should be dried down to a moisture level of 10-12 per cent. Use to 300gauge black polythene lined gunny bags improves the storage efficiency. Store the commodity in wooden boxes at room temperature preferably kept in the curing house for better storage effectively.

Various Methods of Drying

There are several methods that can be used for drying of cardamom. Each method has different advantages and disadvantages.

Sun Drying

Traditionally, cardamom capsules are spread on a concrete floor to dry using the natural heat from the sun. The capsules should be placed away from direct sunlight to preserve the green colour (strong sunlight will make the colour fade). This is the simplest and cheapest method, but does not produce the highest quality product. It is only successful in places where the climate is dry and hot. During the monsoon season for example, drying will be interrupted by rainfall which can cause mould to grow on the capsules. During drying, the capsules may be contaminated by dirt and dust from their surroundings.

Solar Drying

The use of a solar dryer should improve the quality of the dried capsules as it is a cleaner, more controlled environment. However, it is not a popular choice as the green colour is lost during drying. The solar dryer is really only useful in dry hot sunny climates. The capsules should be placed in the dryer, out of direct sunlight, and dried until they have a final moisture content of 10%. In places with high humidity the solar dryer can only be used together with an extractor fan to remove the humid air.

Wood-Fired Dryer

Cardamom capsules are traditionally dried in curing houses, using wood to provide the heat. This method puts a huge demand on firewood. The smoke from the fire can give the capsules an unpleasant smoked flavour. The processor must ensure that the capsules closest to the heat source are not burnt or scorched. Cardamom capsules dried by this method are not of the highest quality.

Electric or Gas Dryer

An electric or gas-fired dryer is an improvement over the use of a wood fuelled fire and is the best choice for drying large quantities of cardamom, especially in places where there is rainfall during the drying season. It is the most expensive of all options but does produce the highest quality product. It is important that the drying temperature does not exceed 50 °C. A range of dryers of different sizes are available depending upon the individual choice and budget. Figure below shows a typical tray dryer.

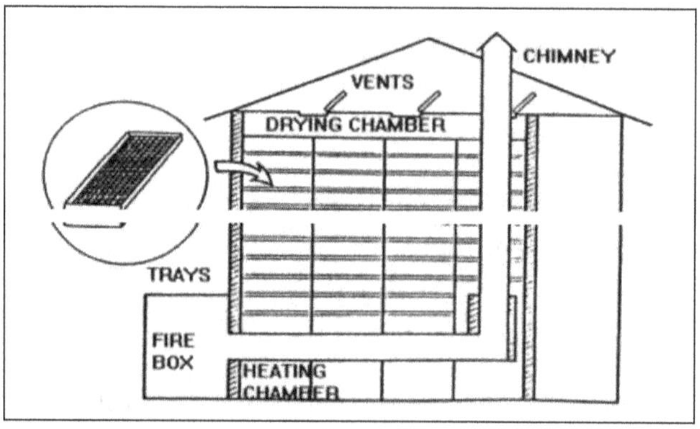

A typical tray dryer.

Humidity-Controlled Drying

A drying chamber has been developed that helps to reduce colour loss and to produce high quality pods. The cardamom capsules are placed in the drying chamber, which is at a temperature of 50 °C. During the first two hours of drying, the humidity builds up within the chamber. This allows the cardamoms to 'cook' and at the same time destroys the enzymes that break down the chlorophyll (chlorophyll gives the pods their green colour). No light is allowed into the drying chamber. After two hours the humid air is blown out of the chamber and the humidity reduced. The capsules are left in the chamber to dry until they have a final moisture content of 10%. Figure below shows a traditional drying chamber.

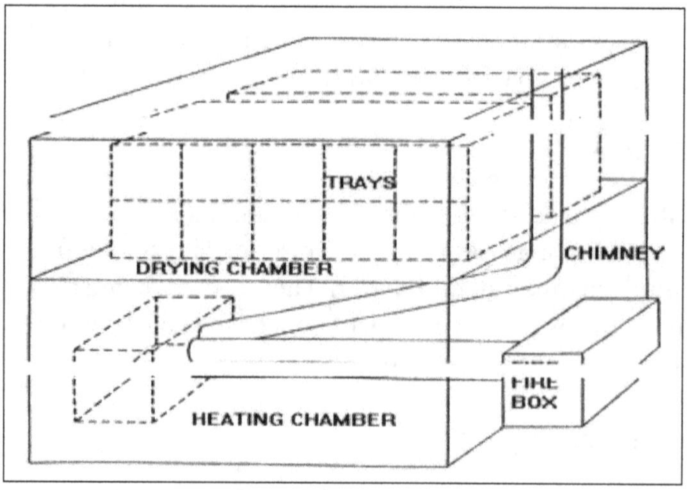

Traditional drying chamber.

The Use of Biomass Gasifiers

Electricity and liquefied petroleum gas (LPG) are clean and convenient fuels for drying, but are not cheap or easily available in villages. Firewood, stubble and dry leaves are readily available in villages, but they are smoky and can contaminate the dried product. A gasifier is a device that has been developed by TERI (The Tata Energy Research Institute) for use in the drying of cardamom. The gasifier uses briquettes that are made from firewood and other types of biomass and turns them into a gas that burns with a clean smokeless flame. The main advantage of using a gasifier is that it is more efficient in terms of the amount of fuel used. Biomass that burns in an open fire loses about two thirds of its energy as smoke. This system therefore uses less fuel and produces a higher quality dried cardamom. The gasifier for drying cardamom, developed by TERI, can be made locally using recycled oil drums.

Drying of Large Cardamom (Amomum Spp.)

Fresh large-cardamom capsules contain about 70±80% moisture (on wet basis) depending upon the maturity level of the capsules at the time of harvest. The whole bunch of cardamom capsules is harvested from august to September, at the end of rainy season but before harvesting the paddy and later each capsule is separated from the bunch/flower and cleaned before drying. In order to achieve a long shelf life and to bring out the aroma, the moisture content has to be brought down below 10% (wet basis) and this process is known as curing, which has to be done immediately after harvesting to avoid fungus infection. In the traditional method, cardamom capsules are spread on a mesh and exposed directly to smoke.

Dry red cardamom capsules.

There is no control on fire and smoke, resulting in poor quality of the dried product. In order to overcome the drawbacks of traditional curing chambers, the Spices Board (a government agency under the Ministry of Commerce) tried to promote the flue-pipe curing chamber generally used for curing of small cardamom. However, the flue-pipe method was found to be slow and also required a very large initial investment, without any energy savings, and has not crossed the demonstration stage as yet. As drying is done during rainy season, peasants consider it generally preferable to smoke the product. 50 kg of seed capsules require 5-7 days for smoke drying. The ratio of fresh weight to dry weight is about 5:1, thus 5kg of fresh harvested capsules will yield 1kg of dry capsules.

Processing of Large-Cardamom in Traditional Bhatti

The fresh large-cardamom capsules contain about 70-80% moisture (on wet basis) depending upon the maturity level of the capsule at the time of plucking. Later on, each capsule is separated from the bunch/flower and cleaned for drying. In order to achieve longer storage life and also to bring out its aroma, it has to be dried to reduce the moisture content to a level below 10% (on wet basis). There are two methods by which cardamom is dried: Traditional method (bhatti) and improved flue pipe curing house.

Traditional Bhatti System

In the traditionally followed system of drying, the fresh cardamom capsules are dried in a locally made curing house commonly known as bhatti. A schematic diagram of the traditional bhatti is given in Figure below showing its structural details and configuration. It consists of a 0:6m thick stone wall structure on all four sides with an opening of about 0:75 m-1:00m in the front to make the _re. On the top of this stone-walled structure a meshed platform, made up of a locally available bamboo mat or chicken wire mesh is placed. It is supported on 0:100{0:150m diameter wood logs. On top of the mesh 0.150{0.200m diameter wood logs are placed on all four sides (cardamom bed area ranges from 4 to 9m2 depending upon the capacity of the bhatti) so that cardamom capsules can be spread within these border logs on the mesh structure.

The mesh is placed about 1.2{1.8 m above the ground level where Fire is located. This height is probably necessitated in order (i) to reduce the charring and burning of cardamom capsules due to flying sparks and (ii) to keep the flue gas temperature below a certain level to prevent over drying. Capsules collected in the estate throughout the day are brought to the curing chamber or bhatti in gunny bags or bamboo baskets. Individual capsules are separated from the bunch, as while plucking the whole capsule bunch is cut from the plant. The fresh capsules are spread in a 200{250mm thick bed on the wire mesh tray by uniformly spreading it with a bamboo or spoon-like wooden spreader. After completion of spreading the fire is ignited.

After completing 20{24 h of drying the capsules are overturned. For this purpose, dried capsules are collected in the bags and partly dried capsules are spread on the meshed structure as bottom layer. Then on this layer the dried capsules are spread. This accelerates the drying process of capsules in the bottom layer and helps in maintaining uniformity in drying operation. After the reshuffling of capsule bed is completed the fire is again accelerated to complete the drying in the next couple of hours. In traditional bhattis the total cardamom drying time varies from 25 to 40 h depending upon several factors such as (i) initial moisture content of capsules, (ii) bhatti structure (opening, mesh, leakage holes, etc.), (iii) _re level maintained (operator tries to maintain low level of fire in bhattis having bamboo mesh and low height taking more time), and also on weather conditions.

Improved Flue Pipe Curing House

About a decade back the Spices Board introduced improved curing houses on a trial basis. The design is similar to the conventional small cardamom curing chambers being used. In this system, there are several wooden trays arranged in stacks. On these wire mesh trays fresh cardamom capsules are spread for drying. Firewood is burned in the furnace and hot flue gas obtained is passed through a network of flue pipes laid down inside the curing chamber. The unit operates on the principle of space heating. The room air surrounding the pipe gets heated up by natural convection and radiation. These systems are of two capacities | 200 and 800 kg. So far, the Spices Board had installed 20 units of 100 kg capacity, 152 units of 200 kg capacity and 1 community-type unit of 800 kg capacity improved cardamon curing houses in Sikkim.

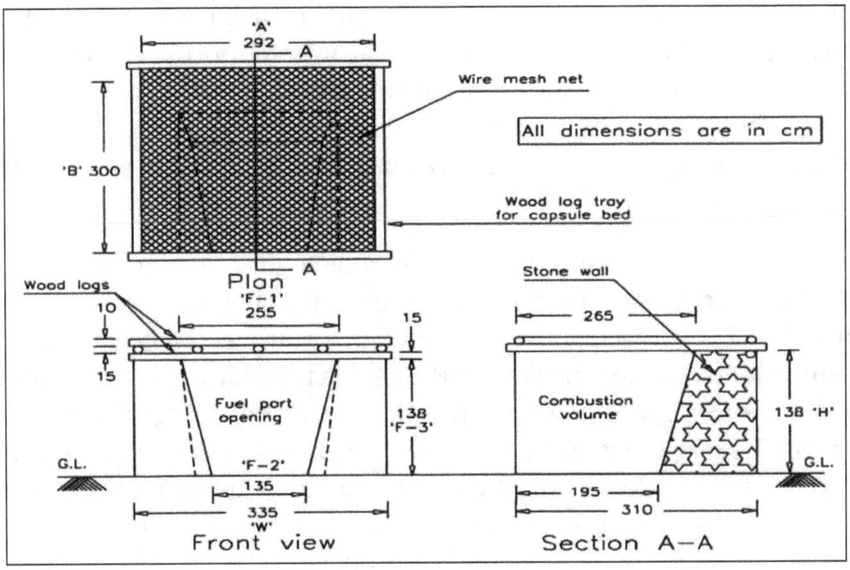

Essential Oils and Oleoresins

Essential oils and oleoresins of spices are considered as luxurious items because of their uses in aromatherapy, food and pharmaceutical industries. Moreover, they also possess antioxidant and antimicrobial efficiency. Cardamom oil is extracted from Elettariacardomomum (also known as Elettariacardomomum var. cardomomum) of the Zingiberaceae family and is also known as cardamomi, cardomum and Mysore cardamom oil properties. Cardamom oil is sweet, spicy and almost balsamic in fragrance, is clear to pale yellow in color and slightly watery in viscosity.

Extraction of Essential Oil

The essential oil of Cardamom is extracted by steam distillation from the seeds of the fruit gathered just before they are ripe. The yield is 1-5%. On steam distillation, the spices yield their volatile constituents. The essential oils thus obtained are endowed with the major part of the spice flavor and fragrance properties. Spice oils, although characterized on the basis of their physicochemical properties, including GLC and spectrophotometric characteristics, are ultimately judged by sensory and olfactory evaluation.

Chemical Composition

The main constituents of its essential oil are Sabinene, Limonene, Terpenene, Eugenol, Cineol, Nerol, Geraniol, Linalool, Nerodilol, Heptenone, Borneol, Alpha Terpineol, Beta Terpineol, Terpinyl Acetate, Alpha Pinene, Myrcene, Cymene, Neryl Acetate, Methyl Heptenone, Linalyl Acetate and Heptacosane etc.

Cardamom Oleoresin

The oleoresins, containing all the volatile as well as non-volatile constituents of the spices, most closely represent the total flavor of the fresh spice; most closely represent the total flavor of the fresh spice in a highly concentrated form. Therefore, oleoresins are preferred spice extract used for flavoring purposes. Oleoresins are the concentrated form of spices where you get the wholesome flavor and aroma of the spice. Oleoresins are characterized by high potency of active components which enables their usage in small dosages.

Oleoresins are preferred because of their microbiological advantages and uniformly in flavor and pungency, easy to store and transport. They have several applications like preparations of beverages, soup powders, curry powders, confectioneries, noodles, canned meats, sauces. Compliance under PFA Act is mandatory. ISI has specified quality standards vide IS 5832 & 7826 of 1975. Oleoresins have large domestic as well as export markets. They are consumed by broad spectrum of manufacturer like beverages, soup powders, curry powders, confectioneries, noodles, canned meats, sauces, poultry products and so on. Most of the end use industries are growing steadily and bound to increase with increase preference of quality products.

Spice Oils and Oleoresins can be used to advantage wherever spices are used, except in those applications where the appearance/filler aspect of spice is important. Usage of spice oleoresins leads to standardization in taste and consistency in flavor. They are also used as a base for a number of seasonings. In green cardamom, greenish to brown semi solid liquid has a great potential to become prime ingredient for not only as food preservative but also flavor enhancer for food industry which is full with triglyceride and steroid components. The product has been very seasonal with the yield mounting up to 4% by using solvent extraction method.

Process Description

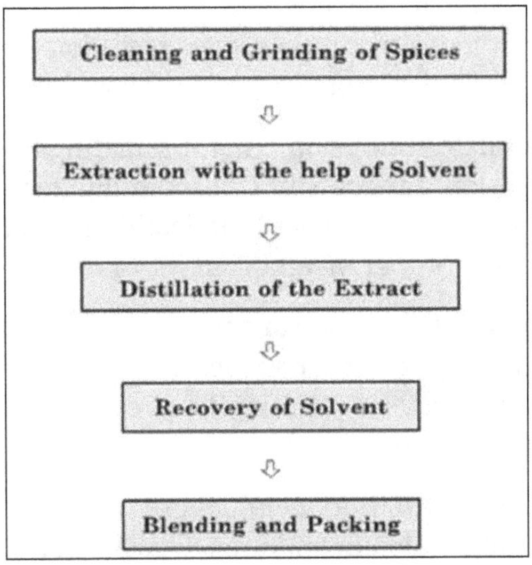

Oleoresins Conversion from Whole Spices.

Various raw spices are cleaned and then ground to require meshing size. The extraction is undertaken with the help of proper solvent. The solvents that can be used are hexane, acetone, ethylene dichloride or alcohol. Extraction is done by percolation of the solvents at room temperature through a bed of ground spices packed in a SS percolator. The dark viscous extract containing not less than 10% of total soluble solids are drawn off and distilled under reduce pressure to remove the excess of solvent.

Health Benefits of Cardamom Essential Oil

The health benefits of Cardamom Essential Oil can be attributed to its properties like anti spasmodic, neutralizes adverse effects of chemotherapy, reduces nausea, anti-septic, anti-microbial, aphrodisiac, astringent, digestive, stomachic, stimulant, diuretic etc. A list of health benefits of Cardamom Oil and Oleorisins:

- Anti-Spasmodic: Cardamom Oil is equally beneficial in curing muscular and respiratory spasms, thereby giving relief in muscle pulls and cramps, asthma, whooping cough etc.
- Anti-Septic and Anti-Microbial: It has very strong anti-septic and anti-microbial properties, which are safe too. If used as a mouth wash by adding few drops of this oil in water, it disinfects the oral cavity of the germs and drives away bad breath. It can also be added to drinking water to kill germs in it. It can also be used in food stuffs as a flavouring agent, which, on the other hand, will keep them safe from spoiling under microbial action. Mild solution in water can be used to bathe to disinfect the skin and hair.
- Digestive and Stomachic: It is the essential oil in Cardamom which makes it a

good digestive. This oil boosts up digestion by stimulating the whole digestive system. It is also Stomachic which means it keeps stomach healthy and functioning properly. It helps maintain proper secretion of gastric juices, acids and bile in the stomach. It also protects stomach from infections.

- Stimulant: It stimulates things up. This stimulating effect also boosts up spirit in cases of depression, fatigue etc. It also stimulates secretion of various enzymes and hormones, gastric juices, peristaltic motion, circulation, excretion etc. thus maintaining proper metabolism.
- Warming: Cardamom Oil has a warming effect. This effect heats up the body, promote sweating, helps clear toughened cough and also gives relief in cold. It also gives relief in headache resulting from cold and used in curing diarrhea caused due to effect of extreme cold.
- Diuretic: It promotes urination, helping lose weight, lower blood pressure, remove toxins and clean calcium and urea deposits in kidneys.
- Aphrodisiac: Cardamom Oil has arousing effects. It helps cure sexual weakness, erectile dysfunctions, impotence, loss of libido or frigidity etc.
- Other Benefits: Cardamom Oil helps neutralize effects of tobacco, insect bites and ingestion of mild poisons, clear bowels, cures colic, removes bad breath, cures oral infections and toothache etc.
- Blending: Cardamom Oil blends well with Rose, Orange, Bergamot, Cinnamon, Cloves, Caraway and Cedar Wood Oil.
- Precautions: No toxic effects have been noticed. Ingestion in overdose may result in unrest and extra heating up of the body and the digestive system, resulting in loose motions and irritation. Pregnant ladies and lactating mothers are advised not to intake Cardamom Oil, since the irritation or warming effect may harm the baby.

Permissions

All chapters in this book are published with permission under the Creative Commons Attribution Share Alike License or equivalent. Every chapter published in this book has been scrutinized by our experts. Their significance has been extensively debated. The topics covered herein carry significant information for a comprehensive understanding. They may even be implemented as practical applications or may be referred to as a beginning point for further studies.

We would like to thank the editorial team for lending their expertise to make the book truly unique. They have played a crucial role in the development of this book. Without their invaluable contributions this book wouldn't have been possible. They have made vital efforts to compile up to date information on the varied aspects of this subject to make this book a valuable addition to the collection of many professionals and students.

This book was conceptualized with the vision of imparting up-to-date and integrated information in this field. To ensure the same, a matchless editorial board was set up. Every individual on the board went through rigorous rounds of assessment to prove their worth. After which they invested a large part of their time researching and compiling the most relevant data for our readers.

The editorial board has been involved in producing this book since its inception. They have spent rigorous hours researching and exploring the diverse topics which have resulted in the successful publishing of this book. They have passed on their knowledge of decades through this book. To expedite this challenging task, the publisher supported the team at every step. A small team of assistant editors was also appointed to further simplify the editing procedure and attain best results for the readers.

Apart from the editorial board, the designing team has also invested a significant amount of their time in understanding the subject and creating the most relevant covers. They scrutinized every image to scout for the most suitable representation of the subject and create an appropriate cover for the book.

The publishing team has been an ardent support to the editorial, designing and production team. Their endless efforts to recruit the best for this project, has resulted in the accomplishment of this book. They are a veteran in the field of academics and their pool of knowledge is as vast as their experience in printing. Their expertise and guidance has proved useful at every step. Their uncompromising quality standards have made this book an exceptional effort. Their encouragement from time to time has been an inspiration for everyone.

The publisher and the editorial board hope that this book will prove to be a valuable piece of knowledge for students, practitioners and scholars across the globe.

Index

www.ingramcontent.com/pod-product-compliance
Lightning Source LLC
LaVergne TN
LVHW080617210926
848485LV00003B/170